40
NIGHTS
TO KNOWING
THE SKY

✦ ✦ ✦

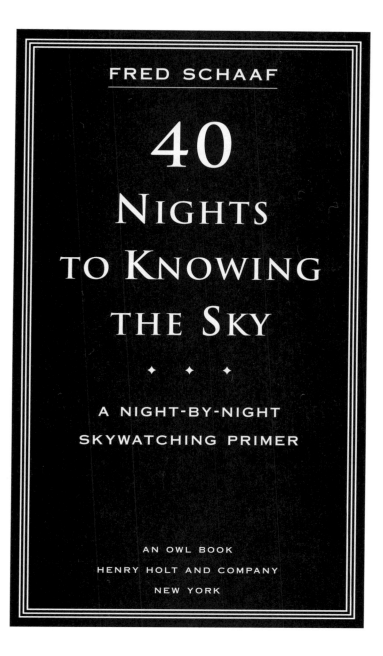

FRED SCHAAF

40
Nights
to Knowing
the Sky

◆ ◆ ◆

A NIGHT-BY-NIGHT
SKYWATCHING PRIMER

AN OWL BOOK
HENRY HOLT AND COMPANY
NEW YORK

Henry Holt and Company, Inc. / *Publishers since 1866*
115 West 18th Street / New York, New York 10011

Henry Holt® is a registered
trademark of Henry Holt and Company, Inc.

Some of the illustrations in this book were adapted with permission
from the work of Doug Myers and Guy Ottewell; the photographs
are reprinted with the permission of Steve Albers and Kosmas Gazeas.
The maps on pages 104–106 are reprinted by permission from
The Universe Explained © 1994 by Marshall Editions Developments Limited.

Library of Congress Cataloging-in-Publication Data

Schaaf, Fred.
40 nights to knowing the sky : a night-by-night
skywatching primer / Fred Schaaf.
p.   cm.
Includes bibliographical references and index.
ISBN 0-8050-4668-2 (pbk. : alk. paper)
1. Astronomy projects—Observers' manuals.
2. Astronomy—Observers' manuals. I. Title.
QB64.S423 1998
520–dc21   97-48507

Henry Holt books are available for special promotions and
premiums. For details contact: Director, Special Markets.

First Edition 1998

Designed by Victoria Hartman

Printed in the United States of America
All first editions are printed on acid-free paper. ∞

5   7   9   10   8   6   4

Contents

✦ ✦ ✦

Part I ✦ Nights of the Heavens in Motion

APPENDICES

ACKNOWLEDGMENTS

✦　✦　✦

I would like to thank David Sobel for his conviction in acquiring this book and working with me through rocky early stages. Jonathan Landreth and Lisa Lester were patient in the further editing of the book.

✦　✦　✦

I wish to dedicate this book to my fellow members of the South Jersey Astronomy Club who have been there at Belleplain State Forest with me on all those "Skywatch" nights when we shared the heavens with each other and with the public. Those nights were an important source of understanding and inspiration for this book.

INTRODUCTION

◆ ◆ ◆

A HUSHED VOICE SPEAKS: *"There's something up there."*

You bet there is. This line has been heard in many a science fiction movie, usually in reference to an extraterrestrial spaceship. But the line is truer than its utterer or the scriptwriter intended. There *is* something up there, and it *does* bring a hush of awe to our voices at times. What is up there is not just something, it's everything: it's the universe we live in. And though that universe may extend to truly inconceivable distances, it is also incredibly close: as close as your backyard, your window, a tilt of your head to the sky.

Welcome to the sky and to your universe. And welcome to a book whose purpose is to take you out on as many as forty clear nights to meet your universe in the most exciting and informative way possible.

Direct observation of the starry heavens is, of course, the most emotionally and intellectually stirring way to learn astronomy. But pure observation needs to be guided by facts and ideas. What is the best way to supply the information and ideas that centuries of scientific study have established, without losing the thrill and joy of seeing the universe "face to face"?

This book attempts a new approach. It is based on the idea that facts and explanations will always be more interesting in response to questions raised by direct viewing of the sky's mysterious objects. In other words, it is when you see something both marvelous and puzzling that your natural curiosity demands an explanation for what you're seeing.

So the book's two central organizing principles are these: (1) to introduce each new *sight* in precisely the order it would be met by a beginning skywatcher outdoors; and (2) to introduce each new *concept* precisely when it is needed to answer a question that a new sight naturally raises in the beginner's mind. Not everyone will experience new sights in precisely the same order. But we can come reasonably close to such an order. Not every new concept can be introduced exactly when it is needed to answer a question about a new sight. But we can come reasonably close to doing so.

And, incidentally, this book is not just for beginners. I believe that even the most expert astronomer will learn something new from trying this method and following this specific order of nights and observations. In particular, I hope that the first third of the book leads its user to a better understanding of the motions of the sky's objects and how they all relate and work together.

✦   ✦   ✦

THE FIRST THREE short chapters or "Nights" familiarize the reader with the fundamental, unmoving locations in the altazimuth system of the sky (horizon, zenith, meridian, lines of azimuth and altitude). They also discuss the features in the night sky that today's beginner is most likely to notice first: the rapidly moving lights of airplanes and satellites, and the prominent blotches of "skyglow" caused by city light pollution. Plenty of additional interesting information—about everything from the size of the sky to twilight sights to UFOs to outdoor lighting fixtures—is provided.

Then, beginning with Night 4, we are off to learn planet identification, the incredible truth behind every sunset and star-rise, and the simple key to knowing where and when and in what phase the Moon will appear tonight.

The book's three major parts reflect three stages in the presentation.

The first part, "Nights of the Heavens in Motion," concerns orienting yourself to the sky and learning the motions (and some other basics) of the most important classes of heavenly objects in the night sky: the Moon, planets, and stars. The crucial instructions for novice skywatchers are how to find—and keep track of—the different objects.

The second part, "Nights of the Heavens in Variety," still deals with naked-eye sights but also fills in the details about the individual planets, stars, and constellations (as well as the Moon and the Sun) and adds a wealth of other phenomena and events, like eclipses, occultations, the brightest deep-sky objects, and meteor showers.

The third major part of the book, "Nights of the Heavens in Depth," introduces you to kinds of heavenly knowledge and splendor that only binoculars and telescopes can reveal. If you don't yet own a telescope, this section should inspire you to obtain one, but it is also true that many of the sights in these final chapters can be glimpsed to some extent by those remarkable, inexpensive and portable "telescopes" that almost anyone can afford to buy or manage to borrow: binoculars.

Throughout the book you'll find some passages italicized. These usually explain how you can make your observations.

Indeed, the entire book is goal-directed and action-oriented. But I have also tried to make every step in the process of learning about the sky and the universe thoroughly enjoyable.

◆   ◆   ◆

IT'S REMARKABLE, but in today's world many people know more about the structure of the universe as revealed by the science of astronomy than they do about the sights in the sky and how these relate to that structure.

For instance, many people know that the solar system consists of moons orbiting planets and the Sun being orbited by planets, asteroids, comets, and countless smaller rocky particles. They know that the Sun is a star, blindingly bright only because it is far closer to us than the other stars. They know that the multitude of stars we see in the night sky are incredibly distant suns, many of which probably have planets of their own. Quite a few people even have heard that billions of stars circle around in a vast congregation called a galaxy (ours is the Milky Way galaxy) and that there are billions of galaxies in this collection of all physical things that we call the universe.

We can get some idea of how large distances are in outer space by resorting to the unit of distance called a "light-year."

A light-year is the distance that light travels in one year. Consider that light (and other wavelengths of electromagnetic radiation, from radio waves to gamma rays and cosmic rays) is the fastest thing known. It travels at about 186,000 miles per second. This means that light can reach the Moon from Earth (or vice versa) in about 1¼ seconds. But the closest approach of the closest planet, Venus, is 100 times farther away. In other words, the light from Venus always takes at least $100 \times 1¼$ seconds = approximately 125 seconds, or just over 2 minutes, to reach us. The light from the Sun takes about 8 minutes to reach Earth. The closest that Pluto ever gets to Earth or the Sun is about 100 times the minimum distance between Earth and Venus. Thus it takes about 100×2 minutes = 200 minutes = over 3 hours for light to reach us from Pluto.

Thus, even Pluto, which currently is about $100 \times 100 = 10,000$ times farther away from us than the Moon, is not much more than three "light-hours" away. How far must a light-year be? About 6 trillion (6,000,000,000,000) miles. And the nearest star system to ours is about 4.3 light-years from Earth. There are individual stars we can see with the naked eye that are 3,000 or 4,000 light-years away. The center of the Milky Way galaxy is about 26,000 light-years distant. The great Andromeda galaxy, which we can glimpse with the naked eye on autumn nights away from urban areas, is almost 3 million light-years distant: the light we see from it tonight left it almost 3 million years ago. Large amateur telescopes can show us galaxies or mysterious quasars (which are possibly superenergetic galactic cores) over 5 billion light-years away; in other words, the light we see tonight left these objects before the Earth and Sun formed. And the farthest reaches of the universe may be something like 15 billion light-years away.

We know of nothing physically faster than light. But thinking about the universe can in a sense bridge the entire universe in an instant. Which is more amazing: the size of the universe, or the fact that the tiny specks in the universe called humans have already learned so much about it?

As wonderful as our ability to learn about the universe is our ability to appreciate the beauty of that universe. And all this learning and appreciation begins with what we see in the night sky itself. This book's first goal is to inspire you to step outside to meet the sights of the night sky—the sights of forty nights. Its second goal is to use that inspiration to help you learn about the workings of the night sky and the greater universe whose face it is. That knowledge will enable you to appreciate the sights in the sky even more deeply and richly.

Something's up there alright. And in the next forty nights you'll encounter a thousand glorious aspects of the night sky and see how they all work together in this masterpiece we call the universe.

PHOTOGRAPHIC AND ARTISTIC ILLUSTRATIONS

✦ ✦ ✦

FIGURES

✦ ✦ ✦

TABLES

✦ ✦ ✦

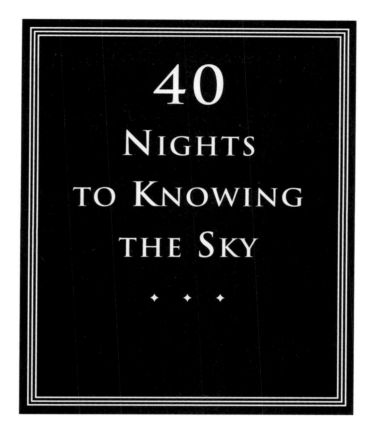

40

NIGHTS
TO KNOWING
THE SKY

✦ ✦ ✦

PART I

NIGHTS OF THE HEAVENS IN MOTION

Comet Hyakutake photographed around 2:30 A.M. on March 27, 1996, by Ray Maher at Jake's Landing in New Jersey. The North Star is the large dot less than an inch to the upper right of the comet's huge head. The author saw the comet's tail stretch almost halfway across the sky at this location and time.

A First Look at the Sky

◆ ◆ ◆

Time: Starting before sunset, ending as night falls.

Place: Anywhere with a mostly unobstructed view of the sky.

IN OUR QUEST to know the night sky, what better way to begin than to follow nature's own entrance into night?

We're all more used to being outside in the day. We're more accustomed to a sky containing Sun and blue, or clearly visible clouds, or some combination of these things. So let's start our observations as the final minutes of familiar day give way to the coming of mysterious night.

LOOKING UP

Sunset is nearing, the sky is clear or partly cloudy. We stand in a place where neither trees nor buildings tower over us in most directions. We are not here to look at the Sun or any clouds. We are here to look at what we so often take for granted: the sky itself.

Look up at the sky.

Those words are as good a magic spell as you're likely to find. If you obey them, if you really stop and gaze for at least a few seconds, there is a strong chance you'll find yourself falling into a genuine state of enchantment. Whatever specific thoughts may go through your mind, one thing is clear: looking at the sky is inspiring.

It can hardly be a coincidence that when our life takes a positive turn, we often say that things are "looking up."

WHAT IS THE SKY, AND HOW BIG IS IT?

But the sky is not just up. *Stare straight ahead. If nothing nearby is blocking your view, you see sky.* The sky is not just over us, but in front of, behind, and to the sides of us. It seems to have the shape of a bowl, half of a sphere.

The bowl of the sky is turned over, with both us and all of the visible world within it. We do not feel confined within it, though—the wind blows around us, we see clouds, which we know to be distant, and the Sun, which we know to be immensely distant. And of course we know that the sky is not really a solid bowl of a particular size at a particular distance above us.

What *is* the sky then, and how big is it?

Big is too small a word to describe the size of the sky! The sky is more than big; it is bigger than anything else we see. In a sense it is bigger than anything else can be. Why? Because the **sky** is simply the entirety of what we can see in directions away from the surface of the planet we live on. By day that means the sunlit atmosphere, the layer of air that extends from ground level with increasing thinness until it grades off into the so-called vacuum of space a few hundred miles out.

Glance over to the setting Sun. Even by day, we can glimpse the Sun and, sometimes, the Moon, and we know that they are beyond our atmosphere, far out in space. They remind us that at night we can see a much darker sky in which we may be able to behold not only the Moon but also planets and stars—many objects at immensely, almost unimaginably great distances.

But we don't have to remember what is visible at night. *We are standing outside with the Sun sinking toward the horizon and night will soon be here for us to experience directly.*

HORIZON AND ZENITH

We say the Sun is sinking toward the horizon. The **horizon,** of course, is the boundary line between sky and Earth. It is the place where the sky seems to meet land or water on the surface of our planet. The horizon marks the edge of the sky—all the way around, on all sides of us.

Take a look now at the horizon around you as night nears. Can you see almost all the way down to the "true horizon" in some directions? Or is your "apparent horizon" much higher—the top edge of a forest (treeline) or a row of buildings (city skyline)?

The **zenith** is the overhead point in the sky. It is as far from the horizon as possible. The zenith is the exact middle of the sky. *If you wish, you can better appreciate that fact right now by lying flat on your back and looking straight up at the zenith and then about at the surrounding sky.*

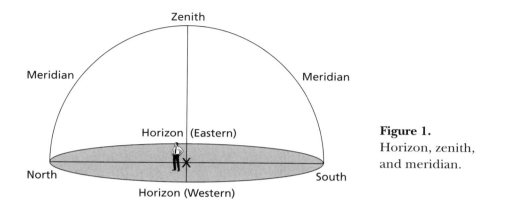

Figure 1. Horizon, zenith, and meridian.

From a standing position, you really have to crane your neck back to look straight up at the zenith.

CENTRAL LINE OF THE SKY

Now that we have names for the edges and the middle of the sky, suppose we want to divide the sky into smaller parts. If our first division is into halves, where do we draw the line that bisects the sky?

The **meridian** is the line that bisects the sky by running from the due north point on the horizon up to the zenith and then down to the due south point on the horizon (see Figure 1). *If you don't know where due north and due south are at the place where you're now standing, just look back toward the setting Sun.* Unless this is the first day of spring or autumn, the Sun will not be setting due west, but at least it will be setting somewhere between southwest and northwest, thus approximately west. And once you know which way west is, you can figure where north and south are and where the meridian runs.

But why is the meridian, dividing the sky into eastern and western halves, more important than a line dividing the sky into northern and southern halves? *To have your answer, you have only to look back at the setting Sun yet again and think where the Sun has traveled during the course of the day.* The answer is that the Sun and almost all heavenly objects—the Moon, planets, and most stars—appear to ascend in the eastern half of the sky and descend in the western half of the sky (we'll learn why in Night 4). Thus, when on the meridian a heavenly object is not just in the middle of its journey across the sky, it is at its highest. And that is when it is typically farthest above any trees or buildings or the dimming haze near the horizon.

When heavenly objects reach this favorable position of being on the meridian we sometimes say that they **culminate.** The Sun, for instance, culminates in the middle of the day.

SUNSET, TWILIGHT, AND NIGHTFALL

At last we come to sunset, and to that complex and magical time between day and night called twilight.

Keep looking at the Sun as it is sinking below the horizon (before that, it is usually dangerously bright). Then look around to all parts of the sky in the minutes after sunset. Try to see which parts of the sky darken first, which stay bright longest, where the gradations from dark to bright are sharpest, whether you notice color not just on any clouds but, usually far more subtly, on different parts of the clear sky itself. Perceive the changes in sounds, smells, and temperature too as night comes on—these are all part of the skywatcher's experience.

If you get cold, tired, or hungry, or have a few chores to do, you might want to go indoors for an hour or more after sunset and then come back out to study the fully darkened sky.

Or you might find yourself standing outside for the whole twilight hour and much longer—simply enthralled with the coming of that half of all nature, that half of all the world, that most of us know so little: the night.

SUMMARY

The sky is the entirety of all that we can see in the universe beyond the surface of the planet we live on. And yet that infinite sky also appears to have a shape, that of a dome, as if we were in the mightiest of all roofed stadiums or cathedrals.

The dome of the sky has an overhead point called the zenith, and a lower edge (the boundary line between Earth and sky) called the horizon. The line that divides the sky into eastern and western halves is called the meridian. When most heavenly objects—the Sun by day, most stars and other objects by night—are on the meridian, they are at their highest.

Sunset gives way to twilight and twilight to night, with many beauties along the way. And we have reached the evening of the first day, the coming of our first night.

✦ NIGHT 2 ✦

MOVING LIGHTS
AND MANKIND'S GLOWS

✦ ✦ ✦

Time: Night (or late dusk).

WE ENDED NIGHT 1'S SESSION with the tranquillity of nightfall.

Now, whether you have decided to stay out later on the same evening and meet the marvels of this second chapter right after the first, or whether this is an entirely different night for you, you will find Night 2's subject moving—literally.

Indeed, in this outing we encounter the night sky's fastest-moving objects—and also its most stationary phenomenon.

MYSTERY OF THE MOVING LIGHT

You are a beginning skywatcher standing outside at night, or possibly late in evening twilight. What is the first thing you notice in the sky, other than its general darkness?

Maybe the Moon (if the Moon is visible at the time).

Maybe lots of stars (if you're far from city lights).

Maybe a stationary point of light here or there, or large regions of glow extending up from various spots along the horizon.

But it's more likely that the first sight that will really stir your interest will be something else: a point of light visibly moving in the sky.

It's a natural human trait to respond to movements, especially rapid ones. And that trait is far more developed in contemporary people who are accustomed to the flashingly fast images of television and other video entertainment. So when people see a point of light in the sky whose motion is immediately detectable, it grabs their full attention.

And the thought that follows, of course, is . . . What is that light?

Is It a UFO? Strictly speaking, every such point of light is at first a UFO—an unidentified flying object. After all, if it is moving through the sky—either Earth's atmosphere or outer space—we could fairly say that it is flying. And until the nature of the object is ascertained, we could fairly say that it is unidentified.

But it's not very likely to be an extraterrestrial spaceship.

Many astronomers do believe in the possibility of extraterrestrial intelligence. But there are plenty of reasons for doubting that a particular point or group of moving lights is something as extraordinary as an alien space vehicle. I'm impressed by the fact that the skywatchers who have spent the most time watching the heavens—thousands of hours!—have rarely, if ever, seen objects that they couldn't identify pretty easily.

We should remain open-minded. But a novice skywatcher can be easily mystified by some of the natural motions of the heavens and by many of the objects and phenomena up there that astronomers already know about.

Airplane, Satellite, or Meteor? Anybody can wildly speculate that a moving light might be an alien spaceship. But it takes a person with a little knowledge to confirm with reasonable certainty whether the moving light is one of three more realistic possibilities: an airplane, a satellite, or a meteor.

We all know what an airplane is, and we know such craft must use lights at night for safety. We'll discuss meteors and artificial satellites, and details of their behavior, in several upcoming Nights (for instance, Night 7). What we want to do now is learn the basics of distinguishing among these three kinds of objects when we see a moving light. The single best initial clue for doing this is apparent speed.

If a point of light streaks across part or most of the sky in somewhere between a fraction of a second and about ten seconds, it is almost certainly a meteor.

If a point of light, or group of lights, takes more than ten seconds but less than several minutes to travel across most of the sky, it is likely to be the light or lights of an airplane.

If a point of light takes several minutes to travel across most of the sky but does noticeably change its position in a matter of seconds, it is likely to be an artificial satellite.

There are many other obvious clues that can assist in distinguishing among these objects. Groups of lights normally belong to airplanes not too distant or too high. Fast moving lights that vanish instantly in the midst of the sky are generally meteors. Satellites visible to the naked eye are plentiful in the few hours after sunset and the few hours before sunrise, but not in the hours of deep night between.

LIGHT POLLUTION

Planes and satellites can be distracting and a bit annoying to those of us who like our nature as pure as possible; they are a major nuisance to astrophotographers. But their light

Satellite image of the USA at night, taken in 1979.

doesn't greatly and constantly reduce the visibility of the stars and the universe. Yet that is exactly what is done by light pollution.

Light pollution is artificial outdoor lighting that is excessive or misdirected. It does no one any good. In fact, it does all of us harm. It wastes energy and money. It works against traffic safety and crime prevention: too much light directly in your eyes can make driving hazardous and detection of a lurking burglar more difficult. It disturbs nocturnal fauna and flora. And light pollution cuts us off from our inspiring natural view of the universe.

The light that shines brightly and directly into our eyes from poorly shielded light fixtures is **glare.** The light that shines uselessly up into the sky and gets scattered back to us from air and any particles in the air is called **skyglow.**

Planes and satellites come and go, but city skyglows unfortunately remain—and maintain the same positions.

LEARN ABOUT OUTDOOR LIGHTING

Take a look around you.

First, note all the individual artificial outdoor lights you can see around you—streetlights, yard lights, business lighting, illuminated advertising signs, building facade lighting, car headlights.

Now look more closely at each artificial light. Is it shielded so that you cannot directly see the source? This is the most important point about outdoor lighting. We want light to shine on the sur-

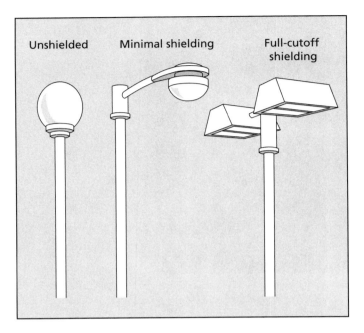

Figure 2.
Different kinds of outdoor light fixtures. Unshielded fixtures like the globe style are most wasteful; full-cutoff fixtures like the shoebox style are least wasteful.

face it is supposed to illuminate (such as a road), not to glare wastefully and dangerously straight into the eyes of motorists or amateur astronomers. If a fixture's shielding permits no light to escape above the horizontal then it is a "full-cutoff" fixture, the kind that is best for most outdoor lighting tasks (see Figure 2).

A second property to note in each outdoor light is the color of the illumination. This is a good clue to what lamp source it uses. Whitish yard lights that are not incandescent light bulbs are typically mercury vapor lights, which are very wasteful and inefficient. Some yard lights and most streetlights are the more efficient yellow (or slightly pinkish-orange) high-pressure sodium (HPS). Some streetlights and security lights are deep gold low-pressure sodium (LPS), which renders almost no color in the objects it illuminates but is extremely efficient (high visibility for low wattage).

There is more to learn about the various kinds of artificial light sources. A good place to turn is the International Dark-Sky Association (IDA). The IDA is a popular clearinghouse for information about outdoor lighting, light pollution, and the growing international campaign to reduce and control light pollution. It has existed since 1988 and has several thousand members. The IDA's web site is given in Sources of Information at the back of this book.

GLOWS OF WASTE

We will take a closer look at how to judge the darkness of the night sky at your location in Night 17. But for now, *look around your observing site and note which regions of the sky are distinctly brighter.*

If you live inside a fairly large city, you will notice that the whole sky has an appreciable brightness to it—maybe it is even a little discolored (probably pinkish)—and that even brighter areas of sky are found extending up from different spots along the horizon. These spots are the regions of the city that throw up the most light pollution, though a fairly small nearby source like a car dealership may, because of its closeness, affect your sky more than a brighter but more distant part of the city.

If you live outside a city, or in a rural area miles away from any city, you will have a darker sky overhead and perhaps darkness low in the sky in one or two directions. But you will notice one or more regions of glow extending up from different points along the horizon. In this case, each patch of brightened sky probably represents the skyglow from an entire city.

SUMMARY

A visibly moving point of light in the sky is almost certain to be the light of an airplane, a satellite, or a meteor. The best initial clue for distinguishing among them is their apparent speed: meteors cross part or all of the sky in less than ten seconds (usually a fraction of a second), airplanes take more like ten seconds to a few minutes, and satellites generally take a few minutes.

Light pollution is excessive or misdirected manmade outdoor lighting. The part of it that goes sideways and shines directly and brightly into the eye is glare; the part of it which goes up and gets scattered back from air and atmospheric particles is skyglow. Except at remote rural locations, there are easily visible patches of skyglow in one or more parts of the sky. Full-cutoff light fixtures provide better visibility with less total light, thus saving energy and money while preserving the darkness of the night sky.

◆ NIGHT 3 ◆

ALTITUDE AND AZIMUTH

◆   ◆   ◆

Time: Any night.

IN THE PREVIOUS NIGHT we encountered moving satellites and meteors, and stationary light-pollution skyglows. But suppose somebody asked you how high in the sky a patch of skyglow extended? Suppose somebody wanted to know the exact direction from which a particular satellite first appeared? Suppose you were curious about how high a star or planet was?

To answer these questions accurately, we need a system of vertical and horizontal measurement in the sky. The simplest such system, the one that takes the dome of sky as its frame of reference, is the **altazimuth system**. The horizon and zenith, which we encountered back in Night 1, can be considered parts of this system.

THE ALTAZIMUTH SYSTEM

Azimuth is horizontal measurement on the dome of the sky. We all know about the cardinal directions—north, east, south, and west—and the points of the compass between them (southwest, north-northeast, and so on). But skywatchers divide the sky more precisely: into a horizontal measure of 360 degrees (360°), with north being 0°, east 90°, south 180°, west 270°, and on to 360°, a figure not used because it is, of course, 0°, due north, to which we have returned. What would be the direction of a star whose azimuth was 85° at a particular time? A little north of due east.

Imagine the lines of azimuth extending from points on the horizon to come together at the zenith (see Figure 3).

Altitude is vertical measurement on the dome of the sky. Let's make a clear distinction here. We are not talking about altitude above Earth's surface measured in feet or miles or kilometers but about altitude measured in degrees above the horizon on the dome of the sky. The observational astronomer deals in **angular altitude**. The degrees are measured from 0° at the horizon to 90° at the zenith. After all, if the sky appears to be a half sphere,

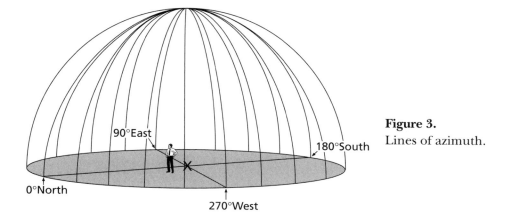

Figure 3.
Lines of azimuth.

the angular distance from one horizon up through the zenith and down to the opposite horizon would be half of the 360 degrees of a sphere or circle, which is to say 180°. And from horizon to zenith would be half of 180°, which is to say 90°.

Imagine the lines of altitude forming concentric circles from the horizon up to the smallest circle (really just a point), the zenith (see Figure 4).

Notice where an airplane sinks below your apparent horizon and try to estimate the azimuth of that point. Or you may find it easier to estimate the azimuth of the Moon or a star at a certain time. You will need to know at least roughly the cardinal directions at your observing site, or use a compass; some compasses even have degrees of azimuth marked on them. *Estimate how high up in angular altitude a skyglow is noticeable or how many degrees high a passing plane gets. Or you may find it easier to estimate the altitude of the Moon or a star at a certain time.* Halfway up the sky (45°)? A third of the way up the sky (30°)?

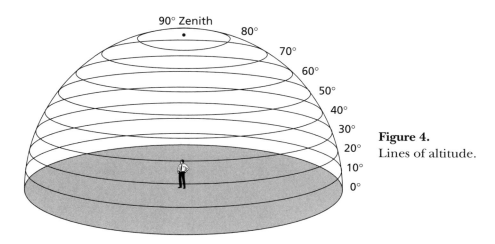

Figure 4.
Lines of altitude.

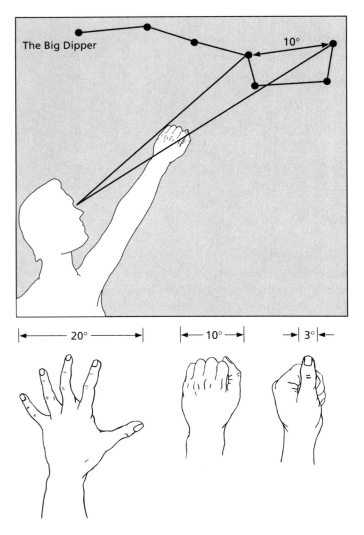

The Big Dipper

10°

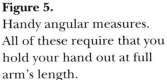

Figure 5.
Handy angular measures.
All of these require that you
hold your hand out at full
arm's length.

|←——— 20° ———→| |←— 10° —→| →|3°|←

A HANDY WAY TO MEASURE IN THE SKY

There is another simple but much more accurate way to measure angular altitude than to just look and try to judge what the fraction of the distance is from horizon to zenith. It involves using an "instrument": your own hand.

Make a fist and then hold it as far out in front of you as you can. Find out how many fist-widths you can fit between horizon and zenith. The answer will be about 9. In other words, the angular width of the fist is about ⅑ of 90°, or 10°. This is true whether you are a small child or an adult basketball player, for the simple reason that small fists tend to come on short arms and large fists on long arms. You can find out more precisely the width of your own fist held out at arm's length by measuring against the known angular length of the Big Dipper (see Figure 5). See Night 12 for information on finding the Big Dipper at any time of night or year.

Your fist, and other arrangements of your hand (see Figure 5), can be used to measure the angular distance between two stars, the angular length of a comet's tail, the angular distance that a meteor traveled. It's not useful for measuring azimuth, though. As Figure 3 shows, the lines of azimuth get closer together as we approach the zenith; therefore, each of the 360 degrees of azimuth must span less sky the higher up we look.

SUMMARY

The altazimuth system consists of a horizontal measure called azimuth (running from 0° at north to 90° at east and so on around through 360°) and a vertical measure called altitude (running from 0° at the horizon to 90° at the zenith). Altitude and the angular distances between objects in the sky can be measured with your fist, which is about 10 degrees wide when seen at arm's length.

✦ NIGHT 4 ✦

BRIGHT STARS AND PLANETS, AND THE TURNING EARTH

✦   ✦   ✦

Time: Any night.

"Horizons roll, and not the standing stars."
—Guy Ottewell

IN NIGHT 2, we learned that any point of light in the sky whose movement was immediately noticeable would be either a meteor (very fast), an aircraft (slow or fast), or a satellite (very slow or slow). An awesome conclusion can be drawn from that discussion: any other speck of light you see above is an object in outer space far beyond our Earth.

But we can easily take the process of identification several steps farther. We can first add the rule that *any bright point of light in the sky whose motion is not quickly detectable will almost always be either a planet or star.* Even in the heavily light-polluted sky of a big city, the brighter planets and brightest stars can be noticed pretty readily with the unaided eye. And there is a way that your unaided eye can easily distinguish for you whether a particular point of light is a planet or a star.

STAR OR PLANET?

Look *around the night sky you are now under. Examine each bright, seemingly unmoving point of light you see up there. Identify whether each is a planet or star by the following simple rule: stars twinkle, but planets shine with a steady light.*

The difference is easily noticeable. I once tested a large group of third-grade schoolchildren to see if they could tell the difference between the twinkling of the star Spica and the steady radiance of the planet Saturn when the two were shining with similar brightness. Out of about three dozen children, essentially all chose correctly which was the star and which was the planet.

On rare occasions, a planet low in the sky might seem to twinkle as much as a star high up. Later in this book we will discuss this and other minor exceptions to the rule, when we examine why turbulence in our atmosphere can shake the image of a star more than that of a planet.

In upcoming Nights, we'll also learn how to recognize the major star patterns and thus identify planets passing through them as interlopers. And you will come to know the planets so well that you will recognize them by other means—including remembering where each is in the current week or season.

RISINGS, SETTINGS, AND THE NIGHTLY JOURNEY ACROSS THE SKY

Whether *the bright specks of light in your sky this evening are all stars, or stars and planets, they all have something in common. Look anywhere in the southern sky and note where a bright star or planet is in relation to a tree or building from a certain location. Then, an hour later, making sure you're standing at the exact same spot, check out the star or planet's position again. You will find that it has moved in relation to the landmark. You will see that it has moved toward the right, which is toward the west.*

What you are seeing is the same behavior that the Sun displays: the Sun rises in the east, travels westward across the sky, and eventually sets in the west. So do most of the stars, the planets, the Moon, and almost everything else in the sky that lies far beyond our atmosphere. Planes and meteors are within Earth's atmosphere, and satellites are on the outer fringes or just beyond, so none of them displays this very slow westward motion.

What causes the **celestial objects**—objects far beyond Earth's atmosphere, in space—to take their nightly (or in the Sun's case, daily) trek from east to west?

THE TURNING EARTH

The answer is well-known, but it is a fact whose full consequences few of us—very few who are not astronomers—have ever thought much about. The celestial objects appear to rise, traverse the sky, and set not because they are really doing so . . . but because the Earth is rotating and carrying us all in the opposite direction.

The wonder of this becomes clear when we put it into the specific situation of you or me out watching the heavens. Most people think of the Earth's rotation as an abstract fact, divorced from everyday life. The best many people can do is to imagine the globe turning in space as it has been pictured before or after the credits of a movie or television show: a ball, usually with no clouds, whirling around incredibly rapidly. At least the turning is displayed as going in the correct direction, eastward. Think a little about this picture and you realize it means that America is heading for the point where Europe is, and will get there faster than a typical jet—to find Europe itself moved onward, of course. To fully appreciate the fact of the Earth's rotation, though, you have to go out and have the concrete experience of watching the Sun, Moon, stars, and planets rise, set, or otherwise move over the course of minutes or hours.

FALLING TO STAR-RISE

Let's try a concrete experiment that demonstrates the reality of the turning Earth directly to our senses. Actually, almost any observation of a celestial object for at least a little while can show us the rotation of the Earth.

Watch a sunset (as we did in Night 1), or a star or planet or Moon rise—or watch a star or planet in relation to a tree or building as suggested above. While watching, remember that the motion you are detecting over the course of several minutes is being produced by the Earth itself rotating you in the opposite direction.

Think what is happening when you watch this fastest apparent motion, the **diurnal motion** (daily motion) of each type of celestial object.

During a sunset, the Sun is essentially standing still while the Earth to the west of us is being turned upward, the lands west of us rising up into a "hill" miles high (and soon miles higher) to block our view of the Sun. When we see what we call a sunset, what we are really watching is an Earthrise—or at least part of the Earth heaving up to the west of us.

When we watch the Moon or a bright planet or star seem to rise in the east, we are really watching that part of the Earth to the east of us sinking to uncover a view of the Moon or star or planet. We're very small on the face of the Earth, and gravity pulls us constantly, firmly toward the center of the planet with everything else we see around us in the landscape and the near sky (clouds and birds and planes), so we don't feel the rapid eastward motion with which the Earth carries us—without friction because the Earth spins in the vacuum of space.

So now you can begin to understand and appreciate the beautiful Guy Ottewell quotation with which we began this Night: "Horizons roll, but not the standing stars." When you observed a bright star or planet poised near a treetop or building in the southern sky, then saw it seemingly further to the right (west) minutes later, the star or planet really never budged. (These objects do move through space, but they are too far away for their own motion to be quickly detectable.) It was the forest, the city skyline, the ridge of mountains rolling to the left (east)—with you, and every person on Earth, being taken along for the ride.

Are there exceptions to this rule that we see the diurnal motion of celestial objects carry them westward? Yes. Notice, for instance, that I required us to look at stars or planets in the southern sky (and for readers in Earth's southern hemisphere, I'd specify the northern sky). But each exception has its reason. And each of these reasons is like a bud, which our examination in the Nights ahead will see open up like a blossoming flower, expanding our awareness of another aspect of how our universe works.

SUMMARY

A bright star can normally be distinguished from a bright planet by its twinkling, which contrasts with the far more steady light of the planet.

When we watch them for a few minutes, most celestial objects—objects far beyond Earth's atmosphere—appear to be moving westward. In reality, this diurnal motion is the result of the Earth rotating us eastward at a relatively rapid rate.

✦ NIGHT 5 ✦

THE MOON AND ITS RACE AGAINST THE TURNING EARTH

✦   ✦   ✦

Time: Any night when the Moon is fairly near a star or planet
(or, any two nights, whether it is near another sky object or not).

IN THE PREVIOUS NIGHT, we saw that Earth's rotation carries us eastward past the Sun, stars, and planets. We saw that since we can't feel the Earth's eastward movement, these bodies look like they are moving westward, rising in the east and traveling across the sky to set in the west. And we saw that although some of these objects are going through space at great speeds, their tremendous distance makes their own motion seem slight, and it is easily overcome by the Earth's fast spinning.

But what about the Moon? As many people know, it is the world closest to Earth in space. It is the first, and so far only world other than our own that human beings have managed to visit in person. From our viewpoint, does Earth's rapid rotation overwhelm the Moon's motion and carry the Moon westward across the sky? Or is the Moon close enough for its true motion in space to be easily detected?

The answer to the first question is yes. But, as we will see, the answer to the second question is also yes.

WHY WE HAVEN'T DISCUSSED THE MOON SOONER

Before we observe the Moon's motion and resolve these questions, you may be wondering why it has taken us until Night 5 to meet with the sky's surpassingly brilliant and prominent object, the Moon. The answer bears directly on the matter of the Moon's motion.

The sky and almost all the sky objects we have already met in this book—planes, meteors, planets, stars, patches of light pollution—can generally be seen at any hour of any clear night. But, despite what many a beginner might think, the Moon is by no means always visible in the night sky. The Moon, in fact, is visible for considerably less than half of the time. If you step outside at night, the Moon will more often than not be absent.

To understand the rules that govern when, where, and in what phase we can see the Moon, we first have to understand the Moon's motion.

THE MOON'S BACKWARD CREEPING

Go out on any night that the Moon appears fairly close to a star or planet. To choose the best time, consult a calendar or almanac and go out on a night when the Moon is between first quarter and full. Carefully measure the distance between the Moon and the other object (use your fist or thumb held out at arm's length to make an angular measurement, as described in Figure 5). Also note the Moon's position (upper left? lower right?) in relation to the star. Then, make the measurement three or four hours later, again noticing the Moon's position with respect to the planet or star.

What do you notice?

Although the Moon's predominant motion seems to be westward with the star or planet, it also seems to be creeping slowly eastward in relation to the star or planet.

Alternatively, note the Moon's position in relation to the straight line between one star that appears above and one star that appears below it. Between the time of your first and second observations, does the Moon cross, or at least move closer to or farther from, this line? Human vision is very sensitive to any departure from a straight line, so you can judge within a matter of minutes when the Moon is exactly lined up with two stars.

Every hour, the Moon moves about one-half degree—one apparent width of itself, one Moon diameter—eastward in relation to background stars or planets. And this is its true motion in its orbit around Earth. There is only one reason we don't see the Moon rise in the west and cross the sky, remaining constantly visible from our half of the Earth for about half a month (half a lunar orbital period) until it sets in the east: the Earth rotates quickly enough to the east to make the Moon seem to fall back and set in the west each day.

THE CAROUSEL AND THE STROLLING MOON

Does all this remain confusing—Earth, Moon, stars, and planets, westward and eastward and backward movement? Then consider the following analogy.

We are on a carousel while somebody is walking very slowly around the carousel in the same direction it is turning. Our carousel, the Earth, takes about twenty-four hours or one day to complete each turn. Meanwhile, the person walking around it—the Moon—takes a bit short of thirty days to stroll all the way around the carousel. Thus, if we stare at the person, we feel as if we are standing still and he (and everything else) is moving backward until he disappears, but each time (each night) we come back around, we see that the person has advanced a little bit farther. Yet how is it that we are judging from our whirling platform that the person has advanced farther? By noticing the person's position in relation to a wall far behind, or in relation to people much farther away—like the planets and stars, which are much farther away than the Moon.

A GOOD CHANCE TO
DETECT THE MOON'S ORBITAL MOTION

When the Moon is quite near a very bright planet or star, its seemingly backward motion can be extremely obvious. This might happen as follows.

1. At nightfall the Moon is in the south, shining to the west of (to the right of) the star.
2. By midevening, although both Moon and star appear to have moved into the southwest (due to Earth's rotation), the Moon is no longer to the right of the star—it is just above or just below it. (It could even pass right in front of the star. See Night 26 for a discussion of such events.)
3. Finally, by midnight, as the Moon and star sink in the west, the Moon is to the east of (left of) the star, and moving away from it.

How will you know which nights are best to see this progression? Exactly which nights the Moon will be passing by various bright stars and planets varies from month to month. You can observe and figure out for yourself a night or two in advance when such a situation will occur. If the Moon is about 12 degrees—a little more than one fist-width—to the right or west of a bright star this evening, it should be quite near the star the following evening. Why 12 degrees? If the Moon's apparent hourly motion in its orbit averages about one apparent Moon-width, about ½ degree, then 24 hours × ½ degree = 12 degrees.

How will you know when the Moon will be in the south (to meet the first requirement of the above progression)? This is easy to predict in advance. The Moon is in the south every month at the same time it is at or near the first quarter phase. As we'll see in our next Night, there is a dependable relationship between the Moon's phases and where and when it appears. You can use knowledge of what the Moon's phase will be to predict exactly where and when to look for the Moon. Conversely, if you know when moonrise or moonset is supposed to occur, you will be able to anticipate which Moon phase you will behold!

THE EASIEST WAY TO DETECT
THE MOON'S ORBITAL MOTION

Through the preceding several paragraphs, we've been assuming we have just one night in which to detect the Moon's eastward motion. But given two nights it's easy to detect the motion. Then it's as if we came around our carousel to sight the walking person again and were immediately struck by the fact that he or she has advanced in relation to the more distant background. *Get a few clear nights in a row and go out at the same time each night. The second night, you will find that the Moon is dramatically farther to the left, in relation both to stars and planets and, you'll see, to any trees or buildings or other objects in your landscape..*

Do you notice anything interesting about the Moon's phase the second night? Or the third night? If you do, you are already at the doorstep of understanding the secrets of the Moon's visibility, the key to an amateur astronomer's life whether he or she observes the Moon or other sights. Our next Night will unlock those secrets almost completely.

SUMMARY

The Moon appears to move westward each night, due to the Earth's eastward rotation. However, the Moon's true motion—the motion of its orbit around the Earth—is *eastward*. The Moon only appears to move westward because the Earth's eastward rotation is so much faster than the Moon's eastward rotation. In order to discover the Moon's true direction, you must compare its position relative to the stars and planets during the course of the night. Careful observation will show that although the Moon, stars, and planets all appear to move across the sky from east to west, the Moon moves to the west more slowly than the stars; in other words, it appears to creep eastward relative to the stars, thus revealing its true motion. This is quite noticeable from one night to the next.

✦ NIGHT 6 ✦

THE WHERE, WHEN, AND PHASE OF THE MOON

✦ ✦ ✦

Time: Any night, at the time of night
when your Moon phase of choice is visible.

MANY PEOPLE actually think that whenever it's night, the Moon must be visible in the sky. When they attend a public skywatching session, they are baffled. They ask, "Where is the Moon?" They don't mean, "Is this a night when the Moon rises late?" What they mean is, "I don't understand this. It's night; how could the Moon not be visible?"

In reality, for any given time of night, the Moon is more often *not* visible. And, for a few days each month—the days around its "new" phase—the Moon is not visible at any hour, in either night or day.

Why is the New Moon invisible? If the Moon rises at 8 P.M. tonight, when will it rise tomorrow night? Which way—right or left (west or east)—does the illuminated half of a first quarter Moon face, and why does the Moon get called first quarter at a time when it is half-lit? Which lunar phase are you likely to see in the daytime? (Did you know that the Moon is often visible in the daytime?) At what times does the Full Moon rise and set?

Amazingly, it is easy to answer all these specific questions—and almost any other concerning the time and place and phase of the Moon—if you understand some simple basics. As a matter of fact, where, when, and in what phase the Moon appears in the sky are really all inseparably related: if you know what the Moon's current phase is, you can figure out immediately when and where it is visible, and vice versa.

THE MOON YOU SEE AT SUNSET

Let's try some observation first. *Go out at sunset on a clear or mostly clear day and look for the Moon. This experiment will be more interesting if you choose a night when the Moon's phase on your calendar or almanac is somewhere between new (or, rather, a few days after new) and full. The Moon after its full and before its new phase is not visible at sunset, so you wouldn't see it at all in the sky. But let's say you do go out at sunset between the dates of New Moon and Full Moon: where do you find the Moon?*

Do answer this question for yourself by actual observation. But as a guideline and assistance to our discussion, here is a rundown of where you *should* find the Moon at sunset.

- Between New Moon and first quarter, the Moon is a silver or smaller-than-half-lit slice—a **crescent** Moon—in the west or southwest.
- At first quarter, the Moon is half lit and in the south.
- Between first quarter and full, the Moon's phase is mostly lit—a **gibbous** Moon—in the southeast or east.
- Finally, at full, the Moon is completely lit and is rising in the east.

These are the combinations of position and phase of the Moon you will always see at sunset: crescent, west or southwest; first quarter, south; gibbous, southeast or east; full, rising in the east. *Go outside and check each of these situations out for yourself.* You'll see that they are true. You could easily memorize these combinations. But at other times of day and night, these phases will appear elsewhere (or be below the horizon altogether). And there are also the phases after Full Moon up until New Moon. You really should not have to memorize combinations of phase and location for every time of day and night. And you need not. You simply need to understand the system that underlies these appearances.

THE MOON'S POSITION IN ITS ORBIT

The key is to understand the positions of the Moon in its orbit in relation to the Earth and the Sun. A diagram displays it all very well.

Figure 6 shows the view if we are looking down from the north, with the Earth at center and the Moon orbiting around it. Rays of sunlight stream from the top of the diagram, and of course the half of the Earth or Moon that is facing away from the Sun is dark, experiencing nighttime. Imagine yourself on the rotating Earth (as you are!). From our view-

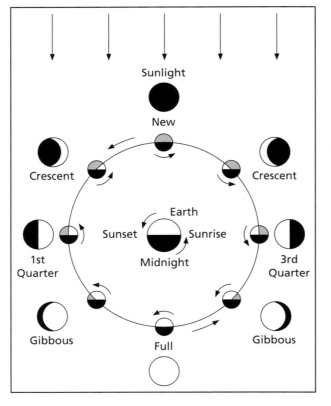

Figure 6.
The Moon's orbital positions and phases. Black represents the night side of the Moon and Earth. Gray represents the sunlit (day) side of the Moon, which is pointed away from the Earth.

point in this diagram the rotation is counterclockwise. When you come to the start of the half of the Earth in darkness, you are experiencing sunset; when you are in the middle of the dark half, you are experiencing midnight; and so on.

Now note in our diagram where the Moon is at different times in relation to an observer on the daytime half of Earth, at the sunset point, at the midnight point, at the sunrise point. Let's study what the observer at sunset sees.

If the observer at the sunset point on Earth looks straight out and sees a Moon high in the sky due south (halfway between rising and setting), what phase does that Moon show? The left half of the side of the Moon facing the observer is hidden from the Sun; it is part of the Moon's night side at this time. The right half of the Moon's face is lit by the Sun. For the observer on Earth, the Sun has just "set," getting blocked from the observer's view by the curve of the Earth to the west—but the Earth is not blocking from the sunlight the part of the Moon that is facing toward the Sun in space. So an observer at sunset who sees the Moon in the due south will see the left (east) half of its face dark and the right (west) half of it bright; it is a half-moon that we nevertheless call first quarter.

Why first quarter? Not because the face of the Moon is one-quarter lit. Our diagram shows clearly that the Moon in this position has completed the first quarter of its monthly orbital journey, which begins at New Moon and ends at the next New Moon. That's why we call it "first quarter Moon." And "third quarter Moon," more often called "last quarter Moon," has completed the third quarter of its monthly journey, or we may say it is beginning the last quarter of its journey. The time of day at which we see last quarter Moon high in the south is the exact opposite of first quarter Moon: instead of at sunset, we see it in the south at sunrise. And although last quarter Moon again presents us with half the Moon's face lit, the lit half is now the left half—the half facing the rising, not the setting, Sun.

FULL MOON AND NEW MOON

Returning to our diagram and the position of an observer at sunset, what can we make of those most important lunar phases of all, Full Moon and New Moon?

When the Moon is full, it is just becoming visible around the curve of the Earth to the observer at the sunset point on Earth—in other words, it is just rising in the east. This arrangement is easy to grasp. The Full Moon is in the opposite direction from the Sun: it rises at sunset, is highest at midnight, and sets at sunrise. When we look toward the rising Full Moon, the Sun is at our backs and the entire face that the Moon shows Earth is also pointed toward the Sun behind us. Full Moon is the time when the Moon's earthward face is experiencing day.

Full Moon, the most visible phase, has its opposite in New Moon, the invisible phase. At new, the Moon is located in the direction of the Sun, and therefore the side of the Moon facing Earth is entirely hidden from the Sun; it is in darkness, experiencing night. The lit (daytime) half of the Moon is facing completely toward the Sun and completely away from us, so there is no part of the lit half available for us to see.

There is another reason why observing the New Moon is normally impossible. Even if there were any part of the lit lunar half to see at New Moon, the Moon would then be too near the glare of the Sun to be seen anyway. New Moon rises almost precisely with the Sun, travels across our daytime sky with the Sun, and sets almost precisely with the Sun.

The Moon's invisibility around New Moon lasts only a few days. Soon the Moon moves slightly away from the direction of the Sun, and we begin to see a thin curve of light, a slender sliver of the Moon's daytime or Sun-facing side. Thus a crescent Moon appears low in the west soon after sunset a day or two after New Moon.

ABOUT AN HOUR LATER EACH NIGHT

In the previous Night we learned that the Moon's orbital motion is slowly eastward, about twelve degrees per day. Now we can see that this motion carries the Moon from the phase when it lies in the direction of the Sun and is therefore invisible—New Moon—through the growing (usually called **waxing**) crescent to first quarter, onward through waxing gibbous, to Full Moon. An observer going out at sunset to watch this progression over the course of two weeks would indeed see the Moon about twelve degrees farther east each day.

An interesting fact is that this eastward motion means the Moon will set about an hour later each night. Each moonset is approximately an hour later than the previous night's (or day's). The waxing crescent sets no more than a few hours after the Sun, first quarter around the middle of the night, waxing gibbous between midnight and dawn, and Full Moon at sunrise. After Full Moon, as the Moon goes through diminishing (usually called **waning**) gibbous to last quarter and onward through waning crescent to New Moon, we see it *rising* approximately an hour later each night. Full Moon rises at sunset, but waning gibbous later in the evening, last quarter around the middle of the night, waning crescent between midnight and dawn. And, of course, New Moon both rises and sets with the Sun.

ALMOST ALL THERE IS TO KNOW

That's almost all there is to know about the where, the when, and the phases of the Moon—almost. You need not bother learning yet about some of the minor modifications and qualifications that more advanced skywatchers add to the system described above.

The time will come, in Night 30, when we study why the Moon, throughout its entire orbiting of the Earth, keeps almost (but not quite!) the same face—the same hemisphere of craters, mountains, and other features—toward Earth at all times.

In Nights 28 and 29 we will study eclipses of the Sun and Moon and why the former don't occur at every New Moon and the latter don't occur at every Full Moon.

These qualifications and exceptions simply open up more fascinating matters to explore. As the saying goes, the exceptions prove the rule.

SUMMARY

Where and when the Moon appears in the sky depend on its position in its orbit around the Earth in relation to the Earth and the Sun. The phase of the Moon is produced by the amount and part of the Moon's earthward face that is illuminated by the Sun, so it too depends on where along the Moon's orbit the Moon is located in relation to Earth and Sun. New Moon rises and sets with the Sun, pointing toward us the Moon's night side; therefore it is invisible. It is followed during the course of a month by the waxing crescent, first quarter, waxing gibbous, full, waning gibbous, last quarter, waning crescent, and then New Moon to begin the cycle again. Each phase has its own time and place of appearance as the Moon moves slowly eastward in its orbit, rising and setting very roughly an hour later each day.

◆ NIGHT 7 ◆

THE MOTIONS OF ARTIFICIAL SATELLITES

◆ ◆ ◆

Time: Within a few hours after sunset,
or within a few hours before sunrise.

BACK IN NIGHT 2, we established that any light in the sky that moves noticeably in seconds was almost certainly either an airplane light, a satellite, or a meteor. Astronomers don't need to learn more about airplane lights, but satellites and meteors we do need to understand.

THE SECRET OF SATELLITES' MOTION

This Night is an especially appropriate time to study satellites and their motion because it is the final step in a progression we've been following in the past few Nights.

- Far: The stars and planets are so distant that their motion through space is almost completely overwhelmed by the nearby rotation of Earth, which carries us observers eastward and makes the stars and planets appear to move westward each night.
- Closer: The Moon is close enough that, even though Earth's rotation overwhelms its motion through space, we are able to notice in a few hours, and especially from night to night, that its true motion is making it creep eastward in relation to the stars, planets, and objects in our landscape.
- Closest: The satellites are almost all in orbits close enough to Earth for their true and their apparent motion to exceed that of Earth's rotation. In other words, most satellites, traveling in the same direction that the Earth rotates (eastward), are seen to rise in the west and move across the sky to the east, outracing the rate at which Earth is carrying us around with its spinning.

ARTIFICIAL SATELLITES VS. NATURAL SATELLITES

So far, when I've referred to satellites, I've meant objects made by humans that have been launched into space to orbit around the Earth. But astronomers also often call the natural rocky or icy bodies that orbit around planets "satellites"—even though they are more popularly known as moons. If we need to make a distinction, we can call the human-made objects orbiting Earth or other worlds **artificial satellites**, and the natural bodies orbiting Earth and other worlds **natural satellites**.

In practice, we seldom hear artificial satellites referred to as moons, even though it is not incorrect to consider them human-made or artificial moons.

SEE A SATELLITE FOR YOURSELF

Technically, a spacecraft piloted by a person that is orbiting the Earth can be called a satellite. And of course no satellites are more exciting to spot than those that, though they look like mere stars, we know to contain living, thinking, dreaming fellow human beings. What a thrill it is to behold spacecraft like the space shuttle and the Russian space station Mir. The newspapers sometimes—and NASA information sources always—provide details about which nights and which times these vessels are going to be visible as they pass by your location.

An important fact to consider when you look for satellites is that they are only visible when they are in direct sunlight, outside the Earth's shadow. What makes an artificial satellite visible is the sunlight shining off of it, just as we see the Moon by reflected sunlight. A rotating satellite of nonspherical shape may therefore present from minute to minute or second to second a very different amount of surface from which to reflect sunlight and appear to vary dramatically in brightness.

The key questions about satellite visibility are these: Does the satellite pass near enough to your vantage point on Earth for it to be visible above your horizon (that is, not hidden by the curved bulk of the Earth), and, is the satellite in sunlight when it is passing you?

As it turns out, most satellites are only high enough to catch sunlight for a few hours after the Sun sets, or for a few hours before the Sun rises, as seen by viewers on the ground below.

Go out as evening twilight is ending and watch the sky for moving lights. If the light is slow-moving (taking maybe a minute or more to cross most of the sky), appears to be a single light (binoculars can help prove this), and seems not to blink off and on rapidly but to vary, sometimes erratically and usually over a number of seconds, then you are very likely seeing a satellite. If it is coming from the west (or northwest or southwest), this further increases the chance it is a satellite. And, finally, another way of distinguishing a satellite from a high-flying plane is to observe it fading quickly from view in a clear sky when heading away from the direction the Sun is in.

What is happening when a satellite dramatically fades out is that the Sun has finally set as seen from the satellite. Thus often at dusk we watch a satellite head further and further to the east until, as it flies over lands that are deeper into night than we are, the body of the Earth at last cuts the Sun off from it, and the satellite plunges into the Earth's shadow, which on Earth we call night.

OTHER SATELLITE MOTIONS

Not all satellites appear to move eastward across the sky. The exceptions are very interesting. One exception is satellites in polar orbits—that is, satellites traveling roughly from pole to pole and back. A satellite in this kind of orbit can survey all parts of the Earth, all of which eventually turn beneath it. Some of these are bright enough to see with the naked eye.

Another kind of satellite, in contrast, always stays more or less over one spot on the Earth's surface. These **geostationary satellites** are extremely important and numerous. After all, we have great need for communications and weather satellites to stay in one position relative to the Earth's surface, to beam transmissions from the western to the eastern hemisphere, and to keep a constant watch on the Atlantic for hurricanes and the Pacific for typhoons. How do geostationary satellites keep in place? To do so, they must maintain just the right speed, and they do this as a consequence of how high their orbit is. If a satellite travels too slowly, its orbit decays. As it then enters Earth's atmosphere, it experiences friction, becomes incandescent, and falls, either vaporizing entirely or reaching the ground in charred pieces. If a satellite travels too fast, it will escape from Earth's gravity altogether. For a satellite to travel at the right speed to stay in orbit and be geostationary it must orbit about 22,700 miles above the surface. This is quite high, so the geostationary satellites tend to appear faint. Amateur astronomers need good telescopes and detailed star charts to identify them.

GLIDING SATELLITES VS. STREAKING METEORS

We just considered how a satellite that moves too slowly will fall out of orbit, enter Earth's atmosphere, and burn up. But the natural objects that enter Earth's atmosphere and burn up are not usually in orbit around Earth to begin with, and they do not enter slowly. These objects just happen to be crossing Earth's path and run into our atmosphere, not at speeds around 18,000 miles per hour, like typical satellites, but at speeds that can greatly exceed 100,000 miles per hour (for instance, in cases when the object encounters the Earth nearly head-on). These objects are usually tiny bits of stone and/or iron and are called meteors when they streak brightly across our night sky.

Meteors are usually only visible for a fraction of a second, a few seconds at most, and their streak is much faster than that of the typically slow-seeming glide or creep of a satellite. Meteors are keys to understanding our solar system, though, and can present us with an incredible variety of sights, some of them spectacular. We'll look at them and their showers and "fireballs" in greater detail later in this book.

SUMMARY

Artificial satellites are human-made machines or spacecraft launched into orbit around the Earth, or, in a few cases, around other planets. Most of them orbit close enough to Earth so that they outpace Earth's rotation and therefore appear to come out of the west and head east. Some satellites are in polar orbits, and others are geostationary; the latter stay in one spot over the Earth's surface. Satellites are only visible by reflected sunlight. Even at their altitude of typically hundreds of miles above the surface, they eventually pass into the Earth's shadow and fade from view, so most satellites are only seen within a few hours after sunset, or within a few hours before sunrise.

◆ NIGHT 8 ◆

THE SUN'S APPARENT JOURNEY AROUND THE ZODIAC

◆ ◆ ◆

Time: Two consecutive clear nights—and also
two nights a few weeks or months apart.

EVEN IF YOU HAVE never studied astronomy before, you may have heard that there are different stars and constellations in winter than there are in summer. Each season has its own stars and constellations in the evening sky. The noble progression of these can be followed every year. But what is the cause of it? We will learn that its *apparent* cause is the Sun's journey through the zodiac, a band of constellations that encircles the heavens, and that its *real* cause is the Earth's annual circling of the Sun.

FOUR MINUTES EARLIER EACH NIGHT

Before we analyze the situation, let's see for ourselves the apparent progression of the constellations. Actually, if you have looked even casually at a bright star or constellation over the course of a few weeks, you may have noticed that it was appearing farther west if you observed it at the same time on various evenings. What's marvelous is that you can perceive this mysterious motion even from one night to the next, if you perform your observations correctly.

Go out at a particular clock time, after nightfall. Pick out an individual bright star (not a planet!) in the south sky. Now walk around until you find a spot from which the star appears poised at the very tip of a treetop or chimney top or other pointed extension from the landscape. Check the precise time of this observation and make certain you can mark the precise place where you are standing.

The next night, return to exactly the same location five or ten minutes earlier, and observe that the star arrives at the tip of the tree or building about four minutes earlier than it did the previous night. When your watch reaches the time you made the observation on the first night, the star is noticeably farther to the right.

Four minutes (actually a few seconds less than four minutes) earlier each night. That means that after half a month, or 15 nights, the star passes the treetop (or the meridian, or the east horizon in rising, or the west horizon in setting) about one hour earlier: $4 \times 15 = 60$ minutes, which is one hour. Check this by going out to observe your star of choice on two nights about half a month apart. Of course, you can check the figures for any interval of nights between observations, by just multiplying the number of nights by 4 to give you the difference in minutes. Thus, after a month a star passes a given point 2 hours earlier. And 2 hours multiplied by the 12 months of the year equals 24 hours. In other words, after one year, the star is back to the same point, perched on your treetop or chimney at the same exact clock time as it was on that date the previous year.

The Sun's Apparent Journey Around the Starry Heavens

What is the cause of this marvelous behavior of the stars? The fact that the star returns to its original position after exactly one year is a vital clue.

The Earth makes one **rotation** (one complete turn around its axis) in a day (24 hours). The Moon makes one complete **revolution** (orbiting) around the Earth in a month. (Actually the Moon takes a little less time—a **lunar month**—to orbit the Earth than our traditional 30-day or 31-day calendar months.) The Earth makes one complete revolution around the Sun in a year (about 365 ¼ days). What we are seeing when we watch a star arrive at a point over the landscape 4 minutes earlier than the previous night, and when we observe the progression of constellations through the year, is the changing perspective of our observation platform—the Earth—as it orbits the Sun.

But the idea that our whole Earth, the home of humanity and the life we know, could possibly be going around another body in space was so astonishing that few thinkers before Copernicus seriously considered it. And, if we, like ancient people, didn't know that the Earth orbits around the Sun, we would interpret the progression of constellations through the year as the effects of the Sun orbiting the Earth.

Think about it. If we noticed that the stars maintained the same patterns relative to each other, we would probably decide (for this and other reasons) that they must be much farther away than the Sun and planets. If the stars were thus "fixed" in their places, we would suspect that the night-by-night, month-by-month westward progress of the stars was probably an illusion produced by a steady eastward progression of the much closer Sun in front of the constellations. And this seems to be confirmed by a number of other observations, some of which we haven't yet made in this book: for instance, the planets also tend to move slowly eastward through the same series of constellations as the Sun.

One month the constellation Taurus sets with the Sun and so is invisible in the solar glare, while the constellation Gemini (east of Taurus) sets just after the Sun and so is visible low in the west at dusk. But since the stars of Gemini keep setting four minutes earlier each night, after another month or so they are setting too soon after the Sun to be seen—the Sun must be shining among, or rather in front of, the Gemini stars. And yet this must be because our most noticeable frame of reference is moving; the frame of reference we call day and night. The Sun causes day wherever it appears, so for it to appear in one constellation after another during the course of a year must mean that either it or our observation platform—Earth—is moving.

We now know that the latter is true, that the Earth revolves around the Sun. Figure 7 shows this correct arrangement. It shows why in December we see the constellation Gemini high in the midnight sky while the constellation Sagittarius is unviewable in the Sun's glare: in December the night side of Earth is turned toward Gemini, while the Sun is in the general direction of Sagittarius as seen—or rather as not seen, because of the solar glare—from Earth. The diagram shows that the opposite is true in June, when Earth's night side looks toward Sagittarius (Sagittarius is visible at night) and Gemini is on the far side of the Sun from us and hence unviewable.

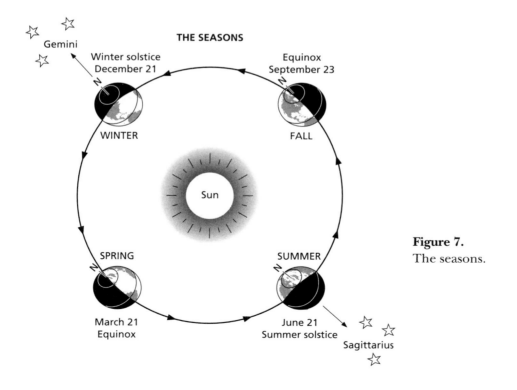

Figure 7.
The seasons.

WHAT CREATES THE SEASONS

Sagittarius is the most southerly constellation of the zodiac, and Gemini the most northerly. And the Sun's southerliness and northerliness are the key to creating the seasons we experience.

In June, the Sun is at its farthest north in the heavens (though from 40 degrees north latitude it passes a little south of the zenith—the farthest north it can get at midday at this latitude is very high in the south half of the sky). In December, the Sun is at its farthest south. But how does this translate into June days being our longest and December's our shortest? It is simply a question of how high the Sun passes. In June, the Sun takes a high arc across the sky, which takes it a long time to complete. In December it travels a low arc across the sky, which takes far less time. Or, we must immediately add, this is true as seen from the middle latitudes of the Earth's northern hemisphere, where most of the world's population lives.

You may have read somewhere that June—the time of the summer solstice—is when the northern hemisphere is tilted the most toward the Sun, and December—the time of the winter solstice—is when the northern hemisphere is tilted the most away from the Sun. This is true, but it is important to understand what it does *not* mean. It does not mean that the tilt of the Earth in relation to the stars changes appreciably in the course of a year. The northern end of Earth's axis keeps pointing toward a place in the sky near Polaris, the North Star. But when the Earth is at the June section of its orbit, the north end of its axis is pointed more nearly toward the Sun (the Sun appears farther north in the sky) than when the Sun is in the December section of its orbit (the Sun appears farther south in the sky). See Figure 7.

THE ECLIPTIC AND THE ZODIAC

Even though we know that the Sun's changing position against the background of stars is caused by the motion of Earth, it is convenient to speak of "the Sun's path among the constellations" and "the band of constellations through which the Sun passes." The former is the **ecliptic.** The latter is the **zodiac.**

The ecliptic is the precise path that the Sun appears to follow in front of the background of stars. We may also think of it as the projection of Earth's orbit onto the sky.

The zodiac is, of course, very famous. It refers collectively to the twelve constellations through which the Sun passes during the year. Actually, according to the modern definition of constellation boundaries, the Sun passes through thirteen—as we'll see later in this book. The zodiac constellations are also the ones in which we generally find the Moon and planets. The reason for this is simple. The ecliptic is the midline of the zodiac and is really the projection of Earth's orbit in the sky. The orbit of the Moon and the planets (except

the oddball Pluto) are in nearly the same plane as that of Earth's orbit. In other words, if we looked at the solar system from a side view, the orbits of the Moon and the planets would be tilted only slightly from that of Earth's. Thus, from within the plane of Earth's orbit, we never see the Moon and planets appear more than slightly above or below the ecliptic in the sky. If we consider the zodiac band to be about 20 degrees wide, centered on the ecliptic, then only Pluto ever appears outside this band of sky and its constellations.

SUMMARY

A star reaches a given position in the sky just under four minutes earlier each night. It returns to the position at the same clock time exactly one year later. What we are seeing is not really a movement of the stars but the effects of our changing position in relation to the Sun—and day and night—as we circle around the Sun in the course of a year. If we rightly consider the stars to be virtually fixed in their positions (because they are so distant), the appearance is that the Sun is moving from one constellation to another during the year, as if it were orbiting the Earth—although we, unlike the ancients, know that really the Earth is orbiting the Sun. The seasons occur because although Earth's axis tilt stays the same (same amount, same direction) throughout the year, this means that the northern hemisphere will be tilted away from the Sun when the Earth is at the December part of its orbit and tilted toward the Sun when the Earth is at the June part of its orbit. The exact path that the Sun takes in front of the background of stars is called the ecliptic, which is actually the projection of our own planet's orbit onto the sky. The band of constellations that has the ecliptic as its midline is called the zodiac.

✦ NIGHT 9 ✦

THE PLANETS' JOURNEYS AROUND THE ZODIAC

✦   ✦   ✦

Time: Any time of night that a planet is visible
(but observe on several different nights
that are a week or even a month apart).

IN THE PREVIOUS NIGHT we saw how the Sun's apparent eastward motion around the zodiac made stars reach a given position about four minutes earlier each night—but that the Sun's seeming motion was really caused by Earth's orbiting of the Sun. At one point in our orbit, Earth's night side faces out to certain constellations, but on the opposite part of our orbit six months later those constellations appear to us in the direction of the Sun and hence are hidden in the day sky.

In this Night, we will see how the planets also seem to be traveling slowly eastward around the zodiac. But in this case the objects really are moving in that direction, moving around their orbits. What complicates their apparent motion in the sky is the fact that we are observing them from one of their fellow planets, which is itself moving slower than some of them but faster than others.

THE PLANETS AS WANDERERS

In Night 4, we learned that many of the planets are bright, some brighter than any star, and that stars twinkle whereas planets typically appear to shine with a much steadier light. Of course, if you have a telescope, you can tell in the most dramatic way possible that what appears to the naked eye as a point of light is a planet: it has a globe, unlike the stars, which are so distant that even in telescopes they still look like points of light.

But we're getting ahead of ourselves in talking about telescopic sights. What other property of planets that can be noticed with the naked eye distinguishes them from stars?

The answer can be found in the very word "planet": it comes from the ancient Greek word *planêtês,* which means wanderer. The ancient Greeks, living long before the invention of the telescope, felt that this was the most important characteristic of the planets: they wandered. The stars keep their positions in their patterns for decades, centuries, even millennia; the shapes of the constellation we call Orion or the star pattern we call the Big Dipper have only changed the slightest bit during the course of human history. But the planets move from one constellation to the next in a matter of weeks, months, or, in the case of the slower planets, years.

In our previous Nights, we maintained that the stars and planets seem to sweep westward during the course of each night because Earth is rotating us eastward; their true motions in space are from our perspective overwhelmed by Earth's rotation because they are so far away. We contrasted planets and stars with the Moon, which, while also seemingly carried westward, is creeping eastward in its orbit enough for this motion to be detected in an hour or so if we can compare the Moon's position relative to a few stars.

Actually, although the planets are hundreds or thousands of times farther away than the Moon, they are still tremendously closer to us than the stars. We, like the ancient Greeks who named the planets "wanderers," can detect the orbital motion of the planets relative to the stars. The true orbital motion of the planets is, like that of the Moon, to the east (left) in our sky. The planets are just so much farther away that the motion appears much slower than the Moon's.

We therefore need to observe the position of a planet on one night and then on another, with the second preferably being many nights later.

The Planets' Race Around their Orbits

I*dentify a planet one night, noting its position in relation to stars near it in the sky. Then, observe where it is in relation to those stars a week and/or a month later. To make these efforts more precise and more tangible, you can sketch or photograph the planet and stars on the several different nights. To photograph, you should have a 35mm camera with a setting (often marked "B") for controlling the shutter manually. Place the camera on a tripod, point the camera at the planet and stars, and use a cable release, fairly fast (high ISO rating) film, and a series of ten- to twenty-second exposures. Which exposure will show the planets and stars best will depend mostly on how dark the night or twilight is and how fast the film is.*

Can you find a planet? If you have trouble, you can check Appendices 8 through 10, which list where to locate various planets in the months and years ahead. (Note, however, that you'll need to read the next few nights—10 through 13—to understand how to use those appendices.)

For detecting the motion of a planet, an important tip is to try to pick a planet that sets long after the end of dusk. A planet that sets soon after dusk may be on the verge of passing near our line of sight with the Sun, and either it or the stars around it might thus become unviewable before you try to make your second observation.

How much of a change in position will your planet make? That depends partly on how far away from Earth and, especially, from the Sun it is. Mercury and Venus can change position rapidly because they are relatively close to the Sun in space, and because Venus can also come closer to Earth than any other planet can. Mars is farther from the Sun than we are, but it too can come fairly close to the Earth and display rather swift movement. On the other hand, Jupiter and Saturn are far away from the Sun and Earth and appear to move quite slowly among the constellations.

The interesting thing is that the farther-out planets like Jupiter and Saturn (as well as Uranus, Neptune, and Pluto) not only have larger orbits, they also travel more slowly in these orbits than the closer-in worlds. The formula that predicts how much more slowly is one of the Three Laws of Planetary Motion, first worked out by the great German astronomer Johannes Kepler (1571–1630). A planet near the Sun, like Mercury, must travel much faster than a farther-out world to keep from being overcome by the Sun's gravity and pulled inward to the Sun.

We can think of the planets circling the Sun as runners around a track. But each "lane" (planetary orbit) is much farther outward than the one before it. Neptune is about 26 times farther from the Sun than Earth is, and so must travel 26 times farther in its outer lane than Earth does in its inner one. It's as if the Earth has to run a mile race in completing one circuit of the track while Neptune has to run a marathon. Yet, according to the Keplerian law of planetary motion we just mentioned, Neptune is also a much slower "runner" than Earth. Instead of taking 26 × 1 year = 26 years to complete an orbit, Neptune actually takes 165 years.

So if you watch for a change in the position of Mercury or Venus in relation to the stars, you'll probably see it in just a few days. But if you watch for Saturn to change position significantly, it might take many weeks. And Neptune—too dim and distant to glimpse without binoculars anyway—takes many years to cross even a single constellation.

STATIONARY POINTS AND RETROGRADE MOTION

But things are by no means as simple as saying the farther a planet is from Earth and Sun, the slower it will always appear to move in the sky.

Sometimes even a nearby planet will—for a while—display a very slow motion in the sky. In fact, you may just happen to observe a planet that seems to have stalled in nearly the same place among the stars for many days; it has reached a **stationary point**.

I wrote above that the true orbital motion of the planets is to the east (left) in our sky. But you might even find that after becoming stationary, a planet moves somewhat west in comparison to the background stars! This is called backward or **retrograde motion**, contrasting with more common, eastward **direct motion**.

Surely these worlds are not really halting in space and moving backward for a while in their orbits? But how could they give the appearance that they are? That is precisely what astronomers in ancient times wondered, and it took until the sixteenth century to start establishing the right answer.

The problem with the ancient scientists' thinking on this mystery was that they started with an incorrect assumption about a crucial point: they assumed that the Sun, Moon, planets, and stars actually all orbited the Earth. They thought that the Earth was the center of the universe.

Our earlier description of how the Moon moves in its orbit was valid because the Earth really is the (approximate) center of the Earth-Moon system. But, as we all know today, Earth is not the center of the solar system. The family of planets and other bodies that make up the solar system do indeed orbit a solar object: the Sun. It was the Polish astronomer Nicolaus Copernicus (1473–1543) who proposed this idea and persuasively explained the planets' motions.

When we consider the motions of the planets, we cannot think of our observing platform, the Earth, as an unmoving vantage point. Earth is moving with all the other planets—slower than Mercury and Venus, but faster than all the others. If the Earth were the center of the solar system, the planets would always appear to move in the same direction in our sky—in the long run, that is: the Earth's rotation would still be making them appear to sweep west across the sky to set each night, of course. But to understand why viewing from an Earth orbiting the Sun causes the planets to appear to halt and go backward (westward) some of the time, let's return to our analogy of the orbital race. Only now, think about just one section of our track, and imagine that its lanes are for driving, not running. The lanes are those of a highway and the Earth is a car we're riding in.

What do we see when we pass a slower "car"—say, Mars—on this highway? At first we see the Mars car moving slowly forward but us getting closer and closer to it. But eventually we see Mars appear to halt and drift backward *as seen against the background of much more distant objects*. In the case of the planets, this background is the stars; if you were driving a car, it would be distant scenery. You'll probably recall having had this experience in real life when you were in one car passing another. If you don't recall your experience that way, study the accompanying diagram (Figure 8). The sightlines in the diagram show clearly that a planet slower than Earth—Mars, Jupiter, and all the still farther out worlds—will indeed appear to move retrograde for a while around the time that Earth is catching up and pulling even with them in the orbital race around the Sun.

So whatever motion—direct, retrograde, or almost none—you have caught a planet performing, you can now understand it. And in understanding it you have a clear glimpse of the reality that eluded thinkers for several thousand years until Copernicus!

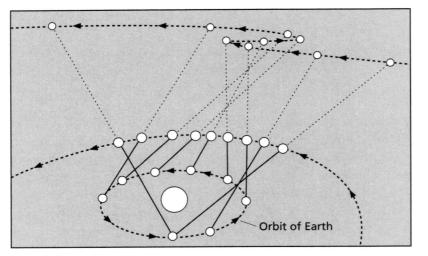

Figure 8.
Retrograde motion
of a superior planet.

SUMMARY

The true orbital motion of the planets in relation to the background stars is eastward in our sky, with planets farther from the Sun moving more slowly in space and most of the time more slowly in our sky. There are, however, times when a planet appears to become stationary relative to the stars, and then appears to move westward with retrograde motion. The appearance of becoming stationary or moving backward—retrograde—is merely an effect of perspective for observers on our moving Earth. What is really happening is that the Earth is either catching up and pulling even with a slower planet, or being caught up and pulled even with by a faster planet.

✦ NIGHT 10 ✦

THE APPARITIONS OF THE SUPERIOR PLANETS

✦ ✦ ✦

Time: The times of night that one or more
of the "superior planets" is visible.

CLASSES OF PLANETS

IN THE PREVIOUS NIGHT, we spoke of planets that were faster or slower than Earth, those that were closer to or farther from the Sun than Earth is. Could we call these two classes of planets "inner" and outer"?

No, it is more useful to take a perspective from outside the solar system and call Mercury, Venus, Earth, and Mars the **inner planets** and Jupiter, Saturn, Uranus, Neptune, and Pluto the **outer planets**. This classification is appropriate because the first four planets are small rocky worlds that all orbit within about the first 150 million miles of the Sun, but are then surrounded by a gap about 350 million miles wide containing little but the sparsely scattered worldlets of the so-called asteroid belt. It's not until we get about 500 million miles from the Sun that we encounter Jupiter, the first of four giant gaseous worlds, each separated by vast distances from the next. Note that Pluto is an outer planet, usually farther out than Neptune, but it is by no means giant or gaseous. It is neither a rocky **terrestrial planet** like Mercury, Venus, Earth, and Mars nor a **Jovian planet**, a **gas giant**, like Jupiter, Saturn, Uranus, and Neptune. We'll speculate about what the oddball Pluto really is later in this book.

If the terms "inner planet" and "outer planet" are thus employed, what terms can we use for the planets closer to and farther from the Sun than Earth is? Astronomers call the former the **inferior planets** (Mercury and Venus) and the latter the **superior planets** (Mars, Jupiter, Saturn, Uranus, Neptune, and Pluto). These terms are, of course, not value judgments; we're not saying that we prefer the superior planets or that they are generally better than the inferior ones! In this usage, "superior" means "greater in distance" (from the Sun than the Earth is) and "inferior" means "lesser in distance" (from the Sun than the Earth is).

The behavior of inferior planets in our sky is so different from that of superior planets that we need to consider the two classes separately, each on its own Night. Our current Night we will devote to the superior planets.

SNAPSHOTS FROM AN APPARITION

The entire progression of different positions that can occur between Sun, Earth, and a superior planet takes over a year (in the case of Mars more than two years). This period of time—how long before the same arrangement of Sun, Earth, and the planet (superior or inferior) occurs again—is called the **synodic period** of a planet as seen from Earth. During part of the synodic period, the Sun gets between Earth and the superior planet and the planet is lost from our view in the solar glare. A planet's **apparition** is the period of weeks or months during which a planet (or other heavenly object) is visible every (clear) night, from when it first appears out of the Sun's glare to when it disappears back into the glare.

Once you understand the progression of positions that occur during a planet's apparition, you'll understand why each major step in the progression features the planet with a different time of rising and setting, different brightness, and—as seen through telescopes—different size and orientation. The progression is majestic and ever fascinating. And I am about to explain how it works.

But this book assumes that you will always be more interested in learning and remembering the hows and whys of the sky if these come as an explanation of some mysterious appearance you have met with outdoors. So how do we encounter the various stages of a superior planet's apparition all in one night? Sometimes it is possible to observe all three of the bright superior planets in one night and use them as examples of different stages in a superior planet apparition.

These views of planets act as "snapshots" (or perhaps we should employ video terminology and say "freeze frames") of an apparition. Appendix 10 lists where and when to look for Mars, Jupiter, and Saturn. With its help you can find the bright superior planets for some years ahead. *Locate and observe, if possible, Mars, Jupiter, and Saturn during the course of a single night. When and where does each appear, when does each rise or set, and how bright does each appear?* After making your observations, check Appendix 9 to determine which of the major positions in an apparition each planet was at or near. The most important positions are called "superior conjunction" and "opposition," and between them occur the times of "quadrature" and the stationary points at which the planet either begins or ends its retrograde motion.

We will now trace verbally the progress of a superior planet through an entire apparition, explaining each of the key positions and describing where and when the planet appears in the sky and what it looks like. Compare your observations with these descriptions.

THE FIRST STAGES OF AN APPARITION

Figure 9 illustrates the arrangement of Earth and the planet for each of the key events in the apparition.

Actually, we seek to trace the planet's course during a full synodic period, and that includes the part before the apparition begins, the part where the planet is too near our line of sight with the Sun to be glimpsed.

When a superior planet is passing due north or due south of the Sun (or, as happens rarely, right behind the Sun), we call this event **superior conjunction**. A **conjunction** occurs when any celestial object passes due north, due south, or, very rarely, directly in front of or behind another in the sky. (In a more general sense, a conjunction is any fairly close pairing of celestial objects in the sky.) When a superior planet has a conjunction with the Sun, it must be a superior conjunction—that is, a conjunction with the Sun that occurs when a planet is farther away from Earth than the Sun is. Inferior planets have both a superior conjunction and an inferior conjunction, the latter occurring on the near side of the Sun.

Superior conjunction is the time when a planet is buried most deeply in the solar glare and is least viewable. But sooner or later the planet that has passed superior conjunction

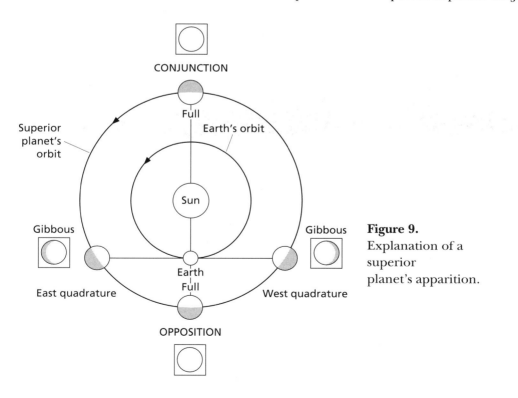

Figure 9. Explanation of a superior planet's apparition.

comes into visibility and begins its apparition for us. Where? Low in the eastern sky, just before sunrise.

The planet tends to be dimmer than usual then because—as Figure 9 shows—it is located on the opposite side of its orbit from us, as far away as it can be except at superior conjunction itself.

Next, we find the planet rising earlier and earlier before the Sun—but immediately a mystery confronts us! The planet seems to be getting farther west with each passing week. But haven't we learned that a planet's true orbital motion is eastward? Could this be a long spell of retrograde motion? No, a careful look at the stars around the planet will indeed show that it is creeping slowly eastward *relative to them*. In other words, although the planet is rising a certain amount of time earlier than sunrise each day, the stars are rising an even larger amount earlier before sunrise each day—so they are moving westward away from the Sun even more rapidly than the superior planet during the same period.

What causes superior planets to rise earlier and earlier in relation to the Sun? The same thing that makes the stars rise earlier each night: the Earth's movement in its orbit, the movement we studied back on Night 8. Earth moves faster than the superior planets, but the eastward orbital motion of a superior planet is able to counter the effect of Earth's motion a little, while the stars can't at all.

THE REST OF THE APPARITION

The superior planet's next major milestone in its apparition is that of **west quadrature**. This occurs when the planet is 90° west of the Sun—halfway across the sky in angular measure. Sure enough, at west quadrature the planet is in the due south, halfway between eastern and western horizons, when the Sun appears on the eastern horizon. The planet now rises around the middle of the night, so we don't have to wake early to view it if we are willing to go to bed late.

Next, the superior planet begins to be overtaken by the Earth, reaching a stationary point in its motion in front of the stars. It then begins to retrograde. The farther out and therefore slower the planet, the sooner the Earth starts to overtake it and the sooner it begins to retrograde. But the planet's westward progress away from the Sun continues whether or not it is retrograding with respect to the stars.

The planet rises earlier and earlier in the evening, brightening as we draw closer to it. Eventually, the superior planet reaches its most favorable position for observation: opposition. **Opposition** occurs when a superior planet is opposite the Sun in the sky and therefore rises at sunset, reaches its highest point in the middle of the night, and sets at sunrise. At opposition, the planet is visible all night long. This is also the time when Earth draws even with it in the orbital race around the Sun and therefore the time when the planet is closest to Earth (see Figure 9). Around the time that the superior planet is closest, it is also brightest and its globe, in telescopes, appears biggest.

After opposition, the planet rises before sunset and is thus already well above the horizon when the sky gets dark enough to see it. As it appears farther west in the sky at each dusk, we go, in reverse order, through the key events of the apparition that occurred before opposition. The planet stops retrograding, reaches its stationary point, and resumes direct motion in relation to the starry background. It comes to quadrature—this time, **east quadrature**—90° east of the Sun, appearing in the due south at sunset and setting around the middle of the night. The planet begins setting sooner and sooner after the Sun. Finally, it is lost from our view low in the Sun's afterglow just after sunset, ending its apparition. Then it returns to superior conjunction, and the cycle starts again.

How Long Is the Synodic Period?

Appendix 6 lists the synodic periods of all the planets, superior and inferior. Jupiter, Saturn, Uranus, Neptune, and Pluto all return to superior conjunction or opposition (or any particular position) somewhere between eleven months and just a few days under twelve months. Jupiter's synodic period is about eleven months, so that it comes to opposition about one month earlier each year. Mars, however, has an enormous synodic period, one which averages 780 days—about two years and two months. The reason it takes so long for Earth to catch back up to Mars for another opposition of Mars is that Mars's orbit is just outside ours and Mars is not that much slower than Earth.

Summary

Planets can be divided into classes such as inner vs. outer and terrestrial vs. Jovian. The planets closer to the Sun than Earth are called inferior planets, and those farther from the Sun are called superior planets. All planets viewed from Earth have a synodic period, the amount of time it takes for them to go through the entire series of their basic positions relative to the Sun and Earth. All planets viewed from the Earth have a period of nightly visibility between one spell of being unviewably close to the Sun in the sky and the next such spell: this period is called an apparition.

A superior planet goes through the key positions of superior conjunction, west quadrature, beginning of retrograde motion, opposition, ending of retrograde motion, east quadrature, and back to superior conjunction. Opposition is the best overall time for viewing a superior planet, for it is then opposite from the Sun in the sky and therefore rises at sunset, is highest at midnight, sets at sunrise, and is also closest, brightest, and biggest as seen from Earth.

✦ NIGHT 11 ✦

THE APPARITIONS OF
THE INFERIOR PLANETS

✦   ✦   ✦

Time: In the hours just after sunset or before sunrise,
when Venus and/or Mercury is visible.

HAVING JUST OBSERVED different positions of superior planets and described the full apparition of one, let us now do the same for the inferior planets, Venus and Mercury.

EVENING STAR AND MORNING STAR

The inferior planets present us with conditions of visibility very different from those of the superior worlds. Venus and Mercury are much closer to the Sun in space than Earth is, and therefore we never see them appear at great angular distance from the Sun in the sky. Mercury can get only 28° and Venus only 47° from the Sun at most; Mercury can rise no more than about two hours before or set about two after the Sun, and Venus can rise only about four hours before or set about four hours after the Sun at most.

Notice that these planets can appear either at dawn or at dusk, but never at midnight. Ancient cultures apparently did not at first understand that the brilliant point of light they would see one season after sunset was the same object as the brilliant point of light that started appearing before dawn a few months later. In some cultures Venus (being far brighter and getting far higher than Mercury) became known at dusk as "the Evening Star" and at dawn as "the Morning Star." Even today, when we recognize that both Evening Star and Morning Star are the same object, and that the object is a planet, we often use these names for their poetic beauty. But it is more common for astronomers to speak of an evening or morning apparition of Venus or Mercury.

VIEWING VENUS AND MERCURY

Using the information in Appendix 8, determine whether Venus is visible this week (it almost always is) and if so whether it currently appears before dawn or after dusk. Then go out and make the easy identification of it: Venus is always by far the brightest naked-eye point of light in the entire heavens. Of all night's usual sky objects, only the Moon is brighter. See how soon before sunrise—or even how long after—you can still spot it, or how soon after sunset—or even how long before—you can spot it.

Also consult Appendix 8 to find whether Mercury is currently visible at dusk or at dawn. In stark contrast to Venus, there are only a few weeks in the year when Mercury is high enough and in a dark enough sky to be easily visible to the naked eye. Mark when the next of these periods is on your calendar and plan for it—unless of course one of those periods happens to be right now, in which case you may be able to observe both of the inferior planets on the same night or even at the same time. Unlike Venus, Mercury can fairly easily be mistaken for a bright star. But most of the time it is the only object bright enough to be seen with the naked eye at its position low in the western dusk sky or eastern dawn sky.

THE EVENTS OF AN INFERIOR PLANET APPARITION

If you observe Venus and/or Mercury this week, what key position in its progression will it be in? The dates these planets reach their key positions in the years ahead are listed in Appendix 8. In order to understand these positions—especially the one you're seeing the inferior planet in right now—let's imitate the example of our previous Night and follow the progress of an inferior planet through an idealized apparition. Or, rather, since one synodic period of an inferior planet includes both an evening and a morning apparition, let's study what happens during an entire synodic period. Inferior planets have both superior and inferior conjunctions with the Sun.

As with the superior planet, a logical place to begin is at superior conjunction. As we learned in our previous Night, superior conjunction is when a planet is in conjunction with the Sun on the far side of the Sun from us. The planet is at its farthest from us then, and because the Sun is approximately between us and it, the planet is lost in the solar glare.

The next step is for the planet to start emerging from that glare in the dusk sky, setting longer and longer after the Sun. This behavior is opposite that of a superior planet, which after superior conjunction emerges from the dawn. A comparison of Figures 9 and 10 shows why. This is the time when we see again the inferior planet, which will eventually overtake the Earth, and the superior planet, which will eventually be overtaken *by* the Earth.

After a while (months for Venus, weeks for Mercury), the inferior planet moves out to its greatest angular separation from the Sun. It appears as far to one side of the Sun as it will ever get from our viewpoint, and roughly as high above the horizon as it will ever get in our before-dawn or after-dusk sky. This position, usually the best overall for viewing an inferior planet, is called **greatest elongation.**

Look at Figure 10. Notice that it shows not only the direction in which the planet appears at major positions in its apparition but also what part of the planet's globe facing us is sunlit as seen through a telescope. That's right: Venus and Mercury show phases in the telescope. The earthward faces of the superior planets can appear only slightly less than fully lit in the telescope because they are so far out that there is little or no difference between which of their hemispheres faces the Earth and which faces the Sun.

I mention these phases here in the naked-eye section of our book because the phase of an inferior planet plays an important role in how bright it appears. After greatest elongation in the evening sky, the planet Venus reaches the point where it has the best combination of apparent size and illuminated amount facing us (the one increases as the other decreases). This is when it shines at its most dazzling, in an event called **greatest brilliancy.**

Back at greatest elongation, a telescope shows Venus or Mercury like little half-moons, their Earth-pointing faces half lit by sunlight. At the greatest brilliancy of Venus, the planet's phase has gotten thinner but it has also gotten closer and thus bigger overall. But now we notice the inferior planet setting rapidly sooner and sooner after the Sun. At last it is a long skinny crescent. The crescent of nearby Venus is so long at this time (Venus is

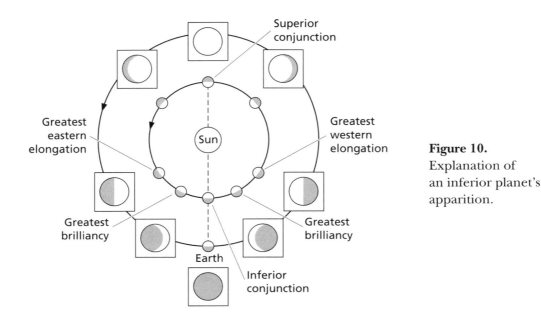

Figure 10.
Explanation of an inferior planet's apparition.

almost at its nearest to Earth) that it can be glimpsed in binoculars and, by sharp-sighted people under very good conditions, even with the naked eye!

Finally, the inferior planet is lost in the solar glare. As it passes between us and the Sun—usually just approximately, not precisely—it reaches **inferior conjunction** with the Sun, a position that superior planets cannot, of course, attain. Venus is so bright and sometimes passes far enough north or south of the Sun that we can actually glimpse it right at inferior conjunction, but this happens rarely.

Next move for an inferior planet? It suddenly appears low in the eastern sky soon before sunrise. Rising progressively earlier than the Sun, it goes through the positions it went through before inferior conjunction in reverse order: greatest brilliancy (of Venus), greatest elongation, and back to superior conjunction, and the synodic cycle begins all over again.

SEASONS OF THEIR VISIBILITY

Mercury and Venus thus appear to vault up in the east before sunrise or in the west after sunset for a certain amount of time before dropping back into conjunction with the Sun. The synodic period of Venus is much longer than that of Mercury. Venus takes about eighteen months to return to the same position with respect to Earth and Sun, and this is divided into an evening apparition and morning apparition of no more than nine months each—actually less, because Venus is not at all easy to see during a spell of several months centered on superior conjunction and during a spell of several weeks centered on inferior conjunction. Mercury has a synodic period of about four months, during which time it may only be visible for a spell of several weeks, or less, centered on each of its greatest elongations.

Not all Mercury's or Venus's greatest elongations are equally good for observers either. The planes of these planets' orbits are not tilted very far from that of Earth, so they do not depart all that far from the ecliptic. But at the middle latitudes of Earth's northern and southern hemispheres, where most of the world's people live, the ecliptic makes a steep angle with the horizon at sunset near the spring equinox and at sunrise near the autumn equinox. At those times, the angular separations of an inferior planet from the Sun are also very steep, meaning that the planet appears almost directly above where the Sun sets or rises rather than appearing slung low to the right or left in the sky. Therefore, Venus and Mercury tend to be highest at greatest evening elongations, which occur within a few months of spring's start, and at greatest morning elongations, which occur within a few months of autumn's start.

SUMMARY

The inferior planets, Venus and Mercury, can never be seen very far from the Sun in the sky. They can rise no more than a few hours before the Sun in a morning apparition, and set no more than a few hours after the Sun in an evening apparition. Venus is the brightest naked-eye point of light in the heavens, usually prominent at either dusk or dawn. Mercury is less bright and usually appears closer to the Sun, and thus is harder to locate and identify. Inferior planets go from superior conjunction to greatest evening elongation to inferior conjunction and then to greatest morning elongation before returning to superior conjunction. Venus is at greatest brilliancy between evening greatest elongation and inferior conjunction, then again between inferior conjunction and greatest morning elongation; these are the times when we get the best combination of the planet being close and its phase not being too skinny.

An evening or morning apparition of Venus can last for most of a year, but those of Mercury are no more than several weeks long. Mercury and Venus are highest at greatest evening elongations, which take place around the start of spring, and greatest morning elongations, which take place around the start of autumn.

✦ NIGHT 12 ✦

CIRCUMPOLAR AND SEASONAL STARS

✦ ✦ ✦

Time: Any night.

IN OUR PREVIOUS three Nights, we explored the motions and special postionings of the planets. But in the night just before these, Night 8, we examined how stars reach a given point in the sky about four minutes earlier each night—and how this leads to a seasonal progression in which certain stars are high in the evening sky in winter, then are replaced by others that are high in the evening sky in spring, and so on through the year.

We decided that it was sometimes more useful to consider the Sun as doing the moving, trekking slowly eastward in the foreground of the zodiac stars. But in reality it was neither the stars nor the Sun that was responsible for the progression. It was our observation platform, the planet Earth, that was moving, orbiting the Sun.

Now that we've reviewed some of our celestial mechanics, it is time to discover some even more interesting things about the apparent motions of the stars in our sky. We will learn, for instance, how it is possible that some stars never rise or set.

THE BIG DIPPER AND THE NORTH STAR

The Big Dipper is not quite the brightest star pattern in the sky; the North Star is outshined by dozens of other stars. Why, then, are they so famous? Or, better yet, why are they the first (and sometimes the only) star and star pattern that a beginning skywatcher is taught? The answer is that they are always visible on a clear night, and they always mark the way to true north. The Big Dipper and the North Star are the most important of the stars that never rise or set when seen from the middle northern latitudes. That is, they are the most important of the north **circumpolar** stars.

Let's find them and see what they do before explaining why they behave as they do.

Go out at night, preferably as soon as darkness falls. If you really don't know which way north is from your observing location, scan around and see if you can chance upon the large, bright seven-star pattern of the Big Dipper. You're better off if you know which way north is. If you do, look to the north sky on any evening of the year and you should be able to find the Big Dipper. The only major problem may be if you are looking on an autumn evening, because then you must have an unobstructed view near the north horizon to spot the Big Dipper. If trees or light pollution prevent you from seeing the low Big Dipper, go out much later in the autumn evening, when you'll find the Big Dipper beginning to wheel up higher into the northeast sky.

Once you have found the Big Dipper, concentrate on the stars that are called "the Pointers." These are the ones on the side of the Big Dipper's bowl that is opposite from the side with the handle. Now draw a line "up" through these Pointer stars—"up" in the direction of the open top of the imaginary pouring utensil. The Dipper as a figure appears upside-down sometimes, in which case the line from the Pointers needs to be drawn down toward the horizon. Extend the line for about one Big Dipper–length. Near the place you are now looking in the sky shines Polaris, better known as the North Star. It actually marks the end of the handle of the Little Dipper. Note that the Little Dipper is not composed entirely of bright stars like its larger counterpart. Only two of its other stars, ones in its bowl, compare with the North Star in brightness.

Now that you have identified the Big Dipper and the North Star, you should never get lost in the dark on a clear night! Polaris marks almost exactly the direction of true north. In contrast, a compass points to magnetic north, which is different relative to true or geographic north from different locations on the Earth—and can vary with Earth's restless magnetic field.

What if forest or clouds hide the north sky? Once you learn most of the constellations, you will find that almost any glimpse of clear sky can provide you with enough information to determine cardinal directions, providing that you know the approximate date and approximate time of night.

CIRCUMPOLAR STARS

The Big Dipper can be used as more than a directional sign to the North Star. It can also be used as the hand of a clock. Figure 11 shows the Big Dipper's orientation and position relative to the North Star at around 9 P.M. for each season of the year. But it also shows this for the Big Dipper at six-hour intervals of a single night and day.

After making your earlier observation, go out later in the night—preferably six hours later—to see where the Big Dipper is. Is the Big Dipper's position the next in the progression of our seasonal diagrams? In other words, is its position consistent with the idea that the Big Dipper completes a full circle, counterclockwise, around the North Star in the course of about twenty-four hours?

Autumn

Winter

Spring

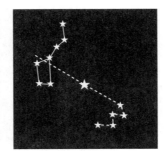

Summer

Figure 11.
Big Dipper, North
Star, and Cassiopeia
at 9 P.M. standard time
in different seasons
(not to scale).

Actually, if you measured carefully, you would find that the Big Dipper gets back to the same position in the circle about four minutes earlier each night—once again, that effect of the Earth changing its position as it orbits the Sun.

But on Night 8, I specified that we look at a star in the southern sky. Most of the stars in the sky rise, cross the sky from east to west, and set. And the time of year when they are well placed for observation, near the meridian or halfway point in the east to west journey, occurs in the evening hours during just one season of the year. Thus we call them the "seasonal constellations." But why are there also north circumpolar star patterns and stars like the Big Dipper and Polaris?

Circumpolar literally means "circling the pole." Polaris is near what we can call the north pole of the sky, the point that lies directly above Earth's geographic North Pole and thus the still point at which the north end of Earth's rotational axis points. Any star or star pattern that is near enough to Polaris in the sky to be able to circle under it without sinking below the due north horizon is north circumpolar.

Polaris's height above your north horizon—and thus how many stars can pass between it and the horizon and be circumpolar—depends on your latitude. If you were at Earth's North Pole (bundled up in a parka, no doubt), you would see the North Star (Polaris) directly overhead. The farther south you went down the sides of the Earth, the lower in the north sky Polaris would shine. By the time you got south to 40° north latitude (the latitude of Philadelphia, Denver, and San Francisco), Polaris would be low enough so that the Big Dipper would just manage to stay above the north horizon as it passed under Polaris. If you live farther south than this, at least part of the Big Dipper ceases to be circumpolar, slipping below the due north horizon for a while each night or day.

Another way to look at this situation is to realize that the seasonal stars that rise and set also are traveling in circles centered on the north pole of the sky (or south pole of the sky, if they are far south enough). The difference is that these stars are far enough from the poles of the sky that their circles are enormous and are cut off by the east or west horizon—so we see them rise and set. But which stars rise and set, which are circumpolar, depends on our latitude. At Earth's North Pole, all visible stars are circumpolar: they travel parallel to the horizon with the North Star at the zenith, the center both of the sky and of their circling. At Earth's equator, no star is circumpolar—unless it is the North Star itself, standing right on the due north point of the horizon, and any star near the south pole of the sky—none is bright, though—standing right on the due south horizon.

SUMMARY

Circumpolar stars are those that are always above the horizon, never rising or setting. The most prominent and famous of the north circumpolar star patterns and stars are the

Big Dipper and the North Star (Polaris). The eye-catching seven bright stars of the Big Dipper include "the Pointers"—the two stars on the far side of the bowl from the Big Dipper's handle. A line drawn through these stars toward the top of the bowl of the imagined Big Dipper figure and then extended about one length of the Big Dipper brings the eye to Polaris, which marks true north. "Circumpolar" literally means "circling the pole." Any star that is close enough to Polaris (the star near the north pole of the sky) to be able to pass below Polaris without sinking below the north horizon for a while is a north circumpolar star. The observer's latitude determines how high in the north sky Polaris appears and how many stars are circumpolar. The entire Big Dipper is circumpolar from about 40° north latitude and anywhere north of there. The Big Dipper makes one full circle around the north pole of the sky and Polaris in about four minutes less than twenty-four hours. Stars too far from the poles of the sky to be circumpolar at an observer's latitude are only high in the evening sky in one season of the year. These seasonal stars move in circles around whichever pole of the sky they are nearest to, but the circles are so huge that they are cut off by the east and west horizons—and we see these stars rise and set.

✦ NIGHT 13 ✦

THE CELESTIAL SPHERE

✦   ✦   ✦

Time: Any night.

WE HAVE COME to the final night of this book's first section. When you have completed this Night you should understand virtually all of the fundamental motions that we notice in the heavens. You should understand the motions of the Moon, Sun, planets, stars, and artificial satellites—not only those caused by the objects' own movements but also those produced by the Earth's rotation around its axis, the Moon's revolution around the Earth, and Earth's revolution around the Sun. You should also understand a host of heavenly appearances: the Moon's phases, the disappearance of most satellites at a certain point in the sky, the greatest brilliancy of Venus, and many more special sights.

Once you have mastered these motions in the heavens, then the rest of this book—despite the riches we have yet to examine—is mostly just descriptive, comparatively easy to understand.

But let's not jump the gun. There is this final Night of the first section. We come to the great encompassing entity known as "the celestial sphere." Among other things, it will help make the previous Night's concepts of circumpolar and seasonal stars at different latitudes, north and south poles of the sky, much easier to understand.

EQUATORIAL SYSTEM AND CELESTIAL SPHERE

On Night 3 we learned the altazimuth system, in which altitude is the vertical measure and azimuth the horizontal. More complex than the altazimuth system, but also more valuable in many situations, is the **equatorial system** of celestial positions.

Look at Figure 12. We imagine that surrounding the sphere of the Earth is a larger sphere, one on whose inside appear all the stars and anything else we can see in the heavens. Even though modern science has revealed that planets, Sun, Moon, stars, and other celestial objects lie at tremendously different distances from Earth, we still feel when we stand outdoors that everything in the sky is at the same distance, everything is attached to the inside of that inspiring dome. But below our horizon, underneath the Earth we stand on, we know there must be the other half of the universe. There are two hemispheres of the heavens enclosing Earth within a **celestial sphere.**

The celestial sphere is a fiction, but a fiction that we find very useful in mapping where to look for celestial objects. We can lay down a gridwork of north-south and east-west lines on the celestial sphere in imitation of the grid of such lines we use to map the Earth. The lines on Earth we call latitude and longitude. We call their counterparts on the celestial sphere **declination** and **right ascension.**

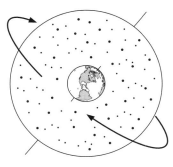

Figure 12.
Earth surrounded
by the celestial
sphere.

DECLINATION

Like latitude, declination (and angular measure in general) is expressed in degrees (°), minutes ('), and seconds (") of arc. The correspondence between latitude on Earth and declination on the celestial sphere is precise and simple, the only difference being the superficial one that a figure for a north or south latitude gets an N or S (for "north" or "south") after it while a figure for a north or south declination gets a + or – in front of it. Thus the city of Philadelphia is near 40° N whereas the star Vega is near +40°—which means it passes virtually overhead as seen from Philadelphia.

This is why the North Star's altitude is always the same amount in degrees as an observer's latitude: if at 40° N the overhead point has a declination of +40°, then it is 50° more down the north sky to the North Star (at about +90°, as Earth's North Pole is 90° N)—and then 40° from North Star to north horizon. Look at Figure 13. Here we see that a person at the North Pole, whose latitude is 90° North, looks straight up to see, at 90° altitude (the zenith), the North Star. Actually the spot on the celestial sphere that is precisely over the north end of Earth's axis is called the **north celestial pole,** and the bright star Polaris lies almost a degree away from it—but that's still very close. In similar fashion, a person on Earth's equator, at a latitude of 0°, would look up to see stars on the halfway line between north celestial pole and south celestial pole pass right overhead and could well understand why this line is called the **celestial equator.** And, as we established in the previous Night, such an observer's view of Polaris would be almost blocked by the bulging Earth to the north; the North Star would sit right on the due north horizon, at 0° altitude.

What about south of Earth's equator? There, the North Star is indeed hidden by the Earth's bulk. Even in those far south lands the rule about Polaris's altitude and the observer's latitude holds, just with a minus sign attached to the altitude: someone at 35° S latitude in Australia would always have Polaris at an altitude of –35°—that is, 35° below the north horizon (hence unviewable). Observers in the southern hemisphere get to see the spot of sky that holds the **south celestial pole** and to see south circumpolar stars circle it. Among the most famous star patterns and stars that are circumpolar for many people at middle latitudes of the southern hemisphere are the Southern Cross and the star Alpha Centauri.

RIGHT ASCENSION

Examine Figure 14. If declination on the celestial sphere is like latitude on Earth then of course right ascension—often called R.A. for short—is like longitude. On Earth, we need a special place to act as the starting point from which to measure east and west longitude.

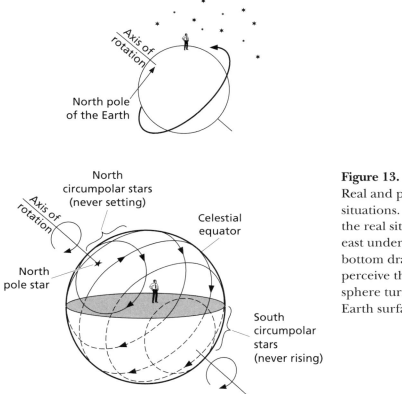

Figure 13.
Real and perceived celestial sphere situations. The top drawing shows the real situation: the Earth turns east under motionless stars. The bottom drawing shows how we perceive the situation: the celestial sphere turns west over motionless Earth surface.

This special place, whose location marks the 0° longitude line, is the Greenwich observatory in England. On the celestial sphere, we choose to start R.A. measurement at the place where the Sun crosses the celestial equator going north. This is the spring equinox point where the Sun lies in the heavens on the first day of spring in Earth's northern hemisphere (the hemisphere in which most of our world's population lives). But there are several differences in the way that R.A. is measured and expressed compared to longitude. For one thing, longitude is measured in degrees east of 0° up to 180° east and degrees west of 0° up to 180° west. But R.A. is measured to the east all the way around the celestial sphere. And it is not usually measured in degrees.

Right ascension is given in 24 "hours" of angular measure, each hour (h) containing 60 "minutes" (m) and each minute containing 60 "seconds" (s). These hours of R.A. begin with the 0h R.A. line, which passes north-south through the spring equinox point, and run eastward all the way around the heavens from there. Thus the famous constellation Orion the Hunter lies mostly between 5h (5 hours) and 6h (6 hours) of R.A., and Sagittarius the Archer mostly between 18h and 19h R.A. Their main patterns lie between about +10° and

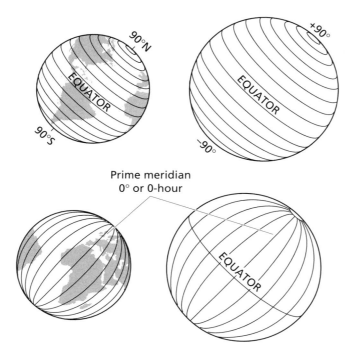

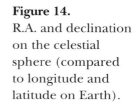

Figure 14.
R.A. and declination
on the celestial
sphere (compared
to longitude and
latitude on Earth).

−10° declination (Orion) and −25° and −35° declination (Sagittarius). If we wanted to be more precise and give the location of a particular star, we could use hours and minutes of R.A. and degrees and minutes of declination. Thus we would say that the star Rigel in Orion is located at 5h 14.5m and −8° 12′.

By the way, because of a very slow wobble in the Earth's axis called **precession** the position of the celestial poles and the whole coordinate system based on them moves, requiring us to specify the "epoch" for which the R.A. and declination of an object is given. The position for Rigel above is for epoch 2000.0, the very start of the year 2000.

But this is a comparatively minor complication for the casual skywatcher. We can truly say that the altazimuth system tells us where in relation to the horizon and zenith, and the cardinal directions, a sky object is for just a brief time—the object's altitude and azimuth are rapidly changing. But the equatorial system gives us a sky "address" where we can find the object in relation to celestial poles and equator night after night—even decade after decade except for the small change caused by precession. *Check the star maps of Figures 15 through 18 (in Nights 15 and 16), and find the following stars: Sirius on the winter map; Spica on the spring map; Altair on the summer map; Fomalhaut on the autumn map. Estimate their R.A. and declination, then check it against the figures for those stars in Appendix 11. Next, go outside at night, recall your latitude, and find the North Star that same number of degrees above the north horizon. Fig-*

ure out how high above the due south horizon the celestial equator should be from your latitude. Look where you estimate this highest part of the celestial equator to be and see if you can find any of the bright stars or patterns that are supposed to be there at this time of year, according to the seasonal maps.

SIDEREAL TIME

A wonderful thing about the equatorial system is that it allows us to use R.A. as the basis for a special time system. Back on Night 8 we saw that a star returned to its same position in the sky every 24 hours minus about 4 minutes. The 24 hours is the length of the solar day, a unit of solar time. But approximately 23 hours and 56 minutes is the length of the sidereal day, a unit of **sidereal time.** Sidereal time means "star time," for the sidereal day is the amount of time it takes for a star to return to the same position in the sky.

When the 5h line of R.A. is on the meridian, we can call this 5:00 (5 hours, zero minutes) sidereal time. When the 12h line of R.A. is on the meridian, we call it 12:00 sidereal time, and of course, the time between the hours can be expressed in hours and minutes (5:36, 12:14, and so on) of sidereal time. What's interesting about sidereal time is that we can consider it independently of either the local time of night or time of the year. It shifts with respect to local time during the course of the year. 5:00 sidereal time occurs around 3 A.M. local standard time in October, 9 P.M. in January, 7 P.M. in February, and 9 A.M. in June. At each one of these local times, the same stars are in the sky in the same positions: Leo the Lion is rising, the Great Square of Pegasus is setting, and Orion is reaching the meridian. But think how different the environments you could observe this sky scene in would be: falling leaves and maybe first frost before an October dawn, bone-chilling cold in the January midevening, snow underfoot after a February dusk, baking heat in the June midmorning. And of course you couldn't see Orion in the blue sky of midmorning (unless there were a total eclipse of the Sun!)—but it would still be there, high in the south, at 5 hours sidereal time.

We could even follow (as I often have) the example of Guy Ottewell and name each one of the hours of sidereal time for the star or star pattern that is then most uniquely and prominently displayed near the meridian. Thus 5h sidereal time could be called "the Orion Hour."

UNIVERSAL TIME

There is another time system that astronomers use that it is important for beginning skywatchers to learn: **Universal Time (UT).** This is the same as civil time (midnight to midnight—though measured in 24 hours, not up to 12 twice like military time) at the 0°

longitude line on Earth, which goes through Greenwich in England. But Universal Time is used as a time independent of the particular astronomical observer's location on Earth.

Here's how to translate Universal Time into your local time if you live in the United States. When you are using standard time (as opposed to daylight savings time) you should subtract the following numbers of hours from Universal Time: in the Eastern time zone, 5 hours; Central time zone, 6 hours; Mountain time zone, 7 hours; Pacific time zone, 8 hours. If the figure you get is a negative number, you must add 24 to it, and the date is that of the previous day; if the figure you get is more than 24 (thinking of the 24-hour clock), subtract 24 to come up with the correct figure for your local time—except that the date is the next day.

What all this means is that the astronomers' day starts at midnight minus 5 hours, 7 P.M. standard time, in the Eastern time zone; midnight minus 6 hours, 6 P.M. standard time in the Central time zone, and so on. Thus if an eclipse is scheduled to begin at, say, 8 UT on January 24, it would begin at 8 minus 5 = 3 A.M. EST (Eastern Standard Time). But if an eclipse was scheduled to begin at 0 UT on January 24, that would be 7 P.M. EST on the previous day, January 23 (0 − 5 = −5 and −5 + 24 = 19 o'clock, which if we translate from 24-hour to 12-hour clock is 7 P.M.).

SUMMARY

The celestial sphere is an imaginary globe that surrounds Earth and includes both the dome of the sky and the equally large dome of heavens underneath us that Earth blocks from view. The stars and other celestial objects are imagined to lie on the inside surface of this sphere. The grid lines of the equatorial system of coordinates used to mark and measure positions on the inside surface of the celestial sphere correspond to (are located directly above) those used on the surface of Earth. There is a celestial equator directly over Earth's equator, celestial poles over Earth's poles; the lines of longitude and latitude on Earth have their counterparts in right ascension (R.A.) and declination on the celestial sphere.

Like latitude, declination is measured in degrees (°), minutes ('), and seconds (") of arc from 0° (equator) to 90° (pole) in each hemisphere, the only difference being that latitude figures are followed by the symbols "N" or "S" for north or south, whereas declination figures are preceded by "+" (plus) or "−" (minus) for north or south. Longitude on Earth is measured 180° west and 180° east from the standard meridian that passes through Greenwich in England. But right ascension is measured in 24 hours of right ascension east from the standard meridian (0 hours R.A.) that passes through the vernal equinox point (where the Sun is in the sky at the start of spring in Earth's northern hemisphere). R.A. figures are given in hours (h), minutes (m), and seconds (s) of arc. Due to the slight wobble of Earth's axis, called precession, the position of the celestial poles and the whole gridwork of R.A. and declination change slightly with time, so the R.A. and declination of a celestial object must be listed for a specific epoch in time (at the time of this book's writing, for the start of the year 2000).

Finally, for some purposes astronomers use sidereal time, measured according to the stars' apparent motions rather than the Sun's. A sidereal day is shorter than a solar day because a star takes only 23 hours and 56 minutes (not a full 24 hours) to return to the same place in the sky. The 4 minutes longer the Sun takes to do so is due, as we saw before, to the fact that Earth is moving along in its orbit around the Sun. When the sidereal time is, say, 9 hours (9:00) the 9h line of right ascension is on the central meridian of the sky (a very useful fact for skywatchers to know). Universal Time (UT) is used by astronomers to provide a single time system independent of the observer's location on Earth. It is based on 24-hour civil standard time (midnight to midnight) of Greenwich in England, so that standard time in the Eastern time zone of the United States (5 time zones west of Greenwich) will, for instance, be 5 hours behind UT.

PART II

NIGHTS OF THE HEAVENS IN VARIETY

♦ NIGHT 14 ♦

THE BRIGHTEST STARS AND MAGNITUDE

♦　♦　♦

Time: A clear night, preferably with little
or no moonlight, haze, or light pollution.

IN THE PRECEDING section of this book we studied the heavens in motion: what causes all the prominent movements of objects in the night sky, how these movements are related to each other, and how they combine to produce the ever-changing drama of the heavens. You now should have an idea of where and when to look for many of the night sky's most important sights, and how to keep track of them once you've found them.

But now that we have the plan for how the heavens work, it is time to begin filling in the details of what is in those heavens—glorious details in astonishing variety.

THE FIRST VARIETY

If we seek to explore the variety of the heavens, we might first mention simultaneously seeing the orb of Moon, the steady-shining planets, and the twinkling stars. But the Moon is not always up, and sometimes our current view may even lack a bright planet. What it doesn't lack, if the sky is not too hazy or moonlit or light-polluted, is stars. The word "astronomy" itself means knowledge of the stars, for they are the most common and integral units of the clear night sky. And the first thing that you'll notice about the stars is that they vary in brightness.

As you gaze around at the stars, how many different levels of brightness do you think you can discern? Roughly how many members of each class of star brightness would you say there are? Selecting only the stars that seem halfway or more up the sky from the horizon, try to rank the brightest stars you see. In a moment, we'll learn where the brightest stars are and what their relative rank is. But first, there is the question of how to classify brightness. Perhaps you chose to recognize dozens of very slightly different brightnesses among the stars in your sky. Or perhaps you bunched them into several large classes of brightness. In a sense, modern astronomy does both.

CLASSIFYING BRIGHTNESS BY MAGNITUDE

Though it has been refined in modern times, the method that the ancient Greek astronomers used still forms the basis by which we classify the brightness of stars and other celestial objects. Those ancient observers categorized stars into six classes of brightness or **magnitude**. The brightest stars were placed in the first class, "first magnitude." The stars of the second brightest class were second magnitude, stars of the third brightest class third magnitude, and so on down to the faintest stars, those of sixth magnitude. Remember that here we mean the faintest stars visible to the naked eye, the unaided eye, for the telescope was not invented and put to astronomical use until much later.

In modern times, astronomers decided that the old magnitude system had to be made more precise and extended to include the brightness of objects much brighter and dimmer than the naked-eye stars. An object one magnitude brighter than another was set as being about 2.512 (actually 2.5118 . . .) times brighter. Why such an odd number? Because this number multiplied by itself 5 times equals 100, so the difference between a star of exactly magnitude 1 and another of exactly magnitude 6 (five magnitudes difference) is 100 times.

Fractions and decimal values of magnitude also came into use. A star could be of magnitude 2.5—midway between stars measured at magnitude 2.0 and 3.0. And it was found that some of the brightest stars were brighter than magnitude 1.0. Some were around zero magnitude. (Remember, the lower the magnitude number, the brighter the object.) The brightest star, Sirius, even carries a negative number magnitude: −1.5. Venus shines at about −4, the Full Moon at about −12.7, and the Sun at a literally blinding −27. On the other end of the scale, it was found that the average person on a clear country night could detect stars as dim as about magnitude 6.5 with the naked eye, binoculars could show tenth magnitude or eleventh magnitude objects, large amateur telescopes could show fourteenth to even sixteenth magnitude objects, and the Hubble space telescope's electronic detectors have recorded some of the most distant objects in the universe at about thirtieth magnitude.

SEEING THE BRIGHTEST STARS

Let's identify the stars that were traditionally considered first magnitude. Today astronomers would technically consider first magnitude stars to be those within 0.5 of magnitude 1—in other words, the stars whose brightness falls between 0.5 and 1.5. But we will here include the splendid stars that are even brighter. What we won't include are any stars not visible from around latitude 40° N. You can get some information about these objects, which include the famous closest star, Alpha Centauri, in Appendix 1, which also provides data on the bright stars we discuss now.

As each star is mentioned, check its position on the star map for its appropriate season (Figures 15–18). Best of all, *observe as many of these stars as you can tonight. The stars listed for each season are those that are well placed, somewhere near their highest in the south sky, at around 9 P.M. standard time (10 P.M. daylight saving time) in that season. But if you stay up until midnight or later, you can get a good view of the bright stars of the next season—for instance, spring's stars if you are observing in winter. In fact, if you go out before dawn on one of the longer nights of the year, you will even start seeing the stars of the second season: summer stars coming up just before dawn in winter.*

THE BRIGHTEST STARS: WINTER

Winter's evening sky presents us with by far the greatest number of first magnitude and brighter stars. Some people think that all the stars shine brighter in winter because of clearer air, but really it's a coincidence that our best collection of brilliant stars is visible then.

Sirius is by far the brightest of all stars and as such is quite unmistakable. Of course, the planets Venus and Jupiter are even brighter, but being planets they generally don't twinkle much. The twinkling or "scintillation" of Sirius is glorious to behold, especially when the star is low and turbulence in Earth's atmosphere can make glints of different color flash in its normally bluish white light. Sirius is the closest star easily visible to the naked eye from around 40° N latitude. It is "only" about 8.6 light-years away. (A light-year, you may recall, is the distance that light, the fastest thing in the universe, travels in one year.) Sirius is located in Canis Major the Big Dog.

To the upper right of Sirius beam two brilliant but contrasting stars, both in Orion the Hunter. The brighter (nearly magnitude 0) is *Rigel,* a star immensely farther away than Sirius and therefore much brighter in reality even if it looks less bright to us. Rigel is a "blue giant" star, tremendously more luminous and huge than our Sun. The other truly brilliant star in Orion is *Betelgeuse* (usually pronounced BET-el-joos by astronomers, though some people jestingly say "beetle-juice"). It is not as luminous as Rigel but is even huger (hundreds of times the diameter of our Sun). It is a "red giant." Its color is more golden orange than red, but is quite noticeable. Betelgeuse is the brightest "red" star in the heavens, and also the brightest to change its brightness noticeably—typically over a period of many months or years.

Aldebaran marks the eye of Taurus the Bull and seems to be part of a V-shaped star cluster that outlines the bull's face. (Actually this light orange star is about twice as close as the other stars of the V.) *Capella,* slightly brighter than Rigel, shines high on winter nights, showing us a slightly yellow tint. It is situated in the constellation Auriga the Charioteer. *Pollux* is the brightest star in Gemini the Twins, but you'll notice quite near it and almost as bright the star Castor, which is just a little too dim for the magnitude 1.50 cutoff figure we use here. *Procyon* is in the small constellation Canis Minor the Little Dog and is almost as close to us in space as Sirius.

THE BRIGHTEST STARS: SPRING

There are three stars that are first magnitude or brighter among the traditional spring stars (Figure 15). Much the brightest is slightly orange *Arcturus*. Extend outward the curve of the handle of the Big Dipper and it brings you to Arcturus, in not very famous Boötes the Herdsman. The saying goes, take an "arc to Arcturus" and then "drive a spike to Spica." Continuing the direction of the arc with a straight line—a "spike"—brings us to *Spica*, the brightest star of Virgo the Virgin. Spica's magnitude is almost exactly 1.0 (Arcturus is a bit brighter than 0.0). Far ahead (westward) of Arcturus and Spica is *Regulus*, marking the heart of Leo the Lion, and famous for having conjunctions with the Moon and planets.

BRIGHTEST STARS: SUMMER AND AUTUMN

Summer has four first magnitude (or brighter) stars of its own. *Vega* is a magnitude 0 blue-white star that passes virtually overhead at Earth's midnorthern latitudes on evenings in the middle of summer. It forms part of the little constellation Lyra the Lyre but is also the peak of the giant unofficial star pattern called the Summer Triangle. The other points of this triangle are first magnitude *Deneb* and *Altair*. Deneb is in Cygnus the Swan and is the most northerly star of the Summer Triangle, passing a bit north of overhead as seen from 40° N in the evenings at summer's end and early autumn. Altair, in Aquila the Eagle, looks brighter than Deneb, but this is partly because it is one of the closest bright stars, just sixteen light-years away. Deneb is a "superluminous" star actually about a hundred times more distant than Altair!

The last first magnitude star of summer is *Antares*, which in many ways is a lesser version of winter's Betelgeuse: its color is similar, but this red giant is less bright. Antares marks the heart of Scorpius the Scorpion prominent but rather low in the south even at its highest.

Even lower (more southerly) than Antares is autumn's only first magnitude star, *Fomalhaut*. This star really stands out in its dim constellation (Piscis Austrinus the Southern Fish) and its entire dimly starred region of the heavens.

SUMMARY

Astronomers measure the brightness of stars and other celestial objects in terms of magnitude. A difference of one magnitude is about 2.512 times, and a difference of five magnitudes is that figure to the fifth power—multiplied by itself five times—or 100. The lower the magnitude number, the brighter the object. The brightest stars, planets, Moon, and Sun all are brighter than either first magnitude or zero magnitude, so negative numbers are used. The brightest stars of winter are Sirius, Capella, Rigel, Procyon, Betelgeuse, Aldebaran, and Pollux. The brightest stars of spring are Arcturus, Spica, and Regulus. The brightest stars of summer are Vega, Altair, Antares, and Deneb. The only first magnitude star among the autumn stars is Fomalhaut.

✦ NIGHT 15 ✦

THE BRIGHTEST CONSTELLATIONS: WINTER AND SPRING

✦   ✦   ✦

Time: Any night, preferably with little or no haze,
moonlight, or light pollution.

"CONSTELLATION" is one of the most beautiful of all words. In Latin it means "a togetherness of stars." In ancient times a constellation could be any pattern of stars. But in the past few hundred years, a consensus has arisen as to which names would be connected with certain star patterns, and in 1930 the International Astronomical Union adopted official boundaries containing the traditional stars of each constellation. Thus, today, a **constellation** is an official pattern of stars or, really, an officially demarcated area of the heavens. **Asterism** is now generally used to mean an unofficial star pattern, like the Big Dipper, which is made up of only some of the stars of the larger pattern and constellation Ursa Major, or the Summer Triangle, which consists of the three brightest stars of three different official constellations.

The system of constellations is a useful one for grasping and remembering where various stars and other celestial objects are located. But in addition, the imaginative figures of the constellations, and the wealth of legend and lore connected with them, are an unending source of delight.

EIGHTY-EIGHT CONSTELLATIONS

There are eighty-eight officially recognized constellations today. One of them, Serpens the Serpent, is composed of two parts—Serpens Caput and Serpens Cauda, Head and Tail of the Serpent—that are completely separated from each other by Ophiuchus the Serpent-Bearer. What was once the largest constellation, Argo Navis (the ship Argo), was broken into three parts, each of which is now an official constellation: Carina the Keel, Puppis the Poop (Deck), and Vela the Sails. Forty-eight constellations were catalogued around A.D. 140 by the astronomer Claudius Ptolemy. Argo was later split in three, for a total of fifty, and the remaining thirty-eight were invented in the centuries since Ptolemy, some of them within the last few hundred years. (The most modern ones are often named after tools and machines of various sorts.) Most of the post-Ptolemaic constellations are composed of stars that were too far south in the sky to be observed properly by the ancient Greeks and Romans.

The largest constellations in the official area covered are, in decreasing size, Hydra, Virgo, Ursa Major, Cetus, and Hercules. The smallest are, in increasing size, Crux, Equuleus, Sagitta, Circinus, and Scutum.

The following subsections of this Night describe briefly the brightest constellations of each season. *Observe as many of the bright constellations of the current season as you can, using Figures 15–18 as your visual guide. What features not described below do you notice?*

BRIGHTEST CONSTELLATIONS: WINTER

Winter not only presents more first magnitude and brighter stars than any other season, but the rest of the constellations in which those stars shine are also bright—rich in second magnitude and third magnitude stars.

Orion the Hunter is the brightest of all constellations. In addition to zero magnitude Rigel and first magnitude Betelgeuse, the main pattern of Orion consists of five second magnitude stars. Three of these stars, of similar brightness and distance apart, form an unmistakable, compact (about three degrees long) northwest-southeast line: the Belt of Orion. From left to right they are Alnitak, Alnilam, and Mintaka. Betelgeuse and Bellatrix are the

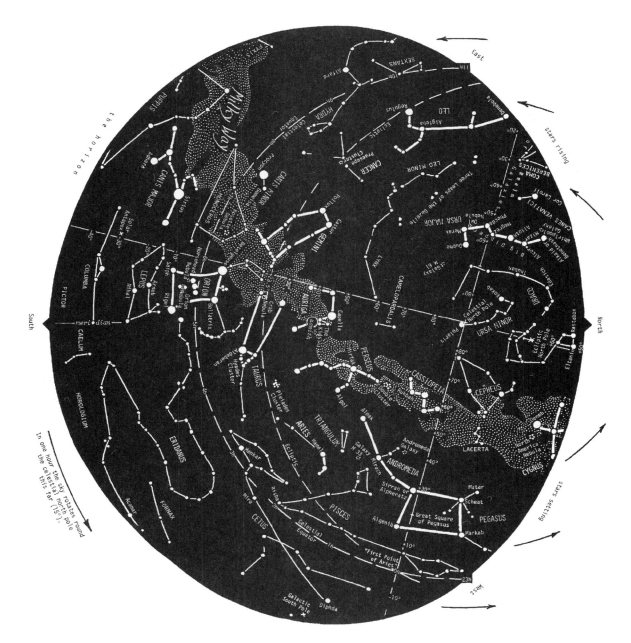

Figure 15.
Winter star map.
Map for January at
about 9 or 10 o'clock
in the evening.

Orion.
(Photographed
by Steve Albers)

shoulders of the giant, Saiph and Rigel his knees. The Sword of Orion is a roughly north-south line of moderately bright stars below the Belt and contains the Great Nebula, which we'll meet on an upcoming night.

Draw a line to the lower left through Orion's Belt, and it will eventually lead you approximately to Sirius, which dominates one of Orion's hounds: *Canis Major the Big Dog*. Canis Major has a number of bright stars, including one, Adhara, that shines at magnitude 1.50 and is thus right on the borderline between a first magnitude and second magnitude star. *Canis Minor the Little Dog* is nothing but bright Procyon and a few dim stars, but Procyon makes it memorable.

Draw a line to the upper right through Orion's Belt, and the line eventually leads you approximately to Aldebaran and the V-shaped Hyades star cluster that together form the face of *Taurus the Bull*. The lovely tiny dipper shape of the Pleiades star cluster is further to the right (west) from Aldebaran, in the shoulder of the Bull. The tips of Taurus's long horns are marked by one star each. The northern horn is tipped with the star Beta Tauri, also known as Nath or El Nath. Interestingly, although this star officially belongs to Taurus, it is also used to help form the neighboring pattern, *Auriga the Charioteer*. Auriga is a pentagon and includes the brilliant star Capella. Capella means "she-goat," and the little triangle of stars near it is called "the Kids"—baby goats that are being held with their mother by the Charioteer.

Winter's final really prominent constellation is *Gemini the Twins*. The twins are two famous heroes from Greek mythology, Castor and Pollux, and the two brightest stars of the constellation, little more than four degrees apart, are named after them. The stars Castor and Pollux—Pollux is slightly brighter, and the southern member of the pair—mark the heads of the Twins. A star not much dimmer than Castor, Gamma Geminorum (Alhena or Almeisan), helps mark the southern feet of the Twins.

BRIGHTEST CONSTELLATIONS: SPRING

Spring has some important constellations that are dim. We'll discover and discuss them on Nights 18 and 20. Here we'll focus on the spring constellations bright enough to be easy to find and see.

High in the south on spring evenings is noble *Leo the Lion.* The front part of the Lion is formed by a backward question mark or hook of stars that is sometimes called "the Sickle." The curve is the head and mane of the Lion, the straight part is his chest. In that chest, the heart (also the handle end of the Sickle pattern) is Regulus. The hindquarters of Leo are formed by a rather bright right triangle of stars quite a distance to the east of the Sickle. The tail end is Beta Leonis, known as Denebola.

High in the north across from Leo is one of the north circumpolar constellations, *Ursa Major the Great Bear.* The famous part of Ursa Major is, of course, the Big Dipper. It forms the hindquarters and unnaturally long tail of the Great Bear. As we saw back on Night 12, spring evenings are the time when the Big Dipper is highest, and that is basically true for the entire constellation. It is also true for the Little Dipper and its constellation, *Ursa Minor the Little Bear.* Since the Little Bear has Polaris at the top of its very long tail, it is always fairly well placed for viewing. But late spring–early summer is when the body or bowl is highest, levitating just above the North Star. It's also when folks in the southernmost United States may glimpse the Southern Cross.

The constellations that hold the stars Arcturus and Spica are not tremendously bright as a whole, so we will actually skip over them for now and mention just one more spring constellation: *Corvus the Crow.* Though its stars are only moderately bright, Corvus is conspicuous because of its compactness: it lies to the southwest of Spica and forms a neat little four-sided pattern.

SUMMARY

A constellation is an official pattern of stars and the area of sky that contains it. An asterism is an unofficial star pattern.

There are eighty-eight official modern constellations. The brightest of winter are Orion, Canis Major, Canis Minor, Taurus, Auriga, and Gemini. The brightest of spring are Leo, Ursa Major and Ursa Minor (both of which are really circumpolar), and Corvus.

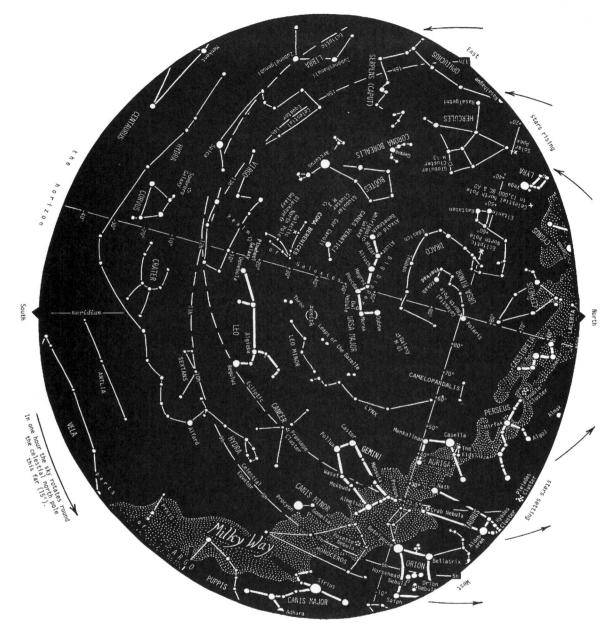

Figure 16.
Spring star map.
Map for April at
about 9 or 10 o'clock
in the evening.

Southern Cross
and Alpha
and Beta
Centauri region.
(Photographed
by Akira Fujii)

✦ NIGHT 16 ✦

THE BRIGHTEST CONSTELLATIONS: SUMMER AND AUTUMN

✦ ✦ ✦

Time: Any night, preferably with little or no haze,

moonlight, or light pollution.

OUR TOUR OF BRIGHT CONSTELLATIONS continues with those of summer and autumn—but first an aside about the naming of stars within constellations.

GREEK LETTER AND OTHER STAR DESIGNATIONS

Several dozen stars have proper names that are used fairly often, and several hundred have proper names that have been fairly well established by scholars though seldom used by amateur astronomers. But the scientific names of the naked-eye stars and some fainter stars includes the Latin genitive (possessive) form of their constellation's name. In other words, the brightest star in Leo the Lion has the proper name Regulus, but its scientific name is Alpha Leonis—the alpha star of Leo. Alpha is the first letter of the Greek alphabet and, ideally, would be assigned to the brightest star of each constellation, with the second brightest star being given the second letter, beta, and so on down to the last letter and faintest star. In reality, Johann Bayer and other astronomers who assigned these scientific names used several different criteria for applying the Greek letters. For instance, in some cases, the alpha star is the first on one end of the constellation, followed by beta, gamma, and so on, according to position. Nevertheless, in many constellations the alpha star is the brightest and beta the second brightest.

But constellations usually contain far more naked-eye stars than letters in the Greek alphabet. So makers of star catalogues after Bayer numbered stars in order of increasing right ascension (thus west to east) within a constellation, or used letters of the modern alphabet or other designations. At first these were still linked with the genitive form of the constellation's name, though. So we have stars with scientific names like "43 Leonis" (43rd star of Leo), 74 Geminorum (74th star of Gemini), and G Scorpii (the G star of Scorpius—which at magnitude 3.2 may be the brightest star not to have a Greek letter designation).

Thousands of dimmer stars visible in binoculars and small telescopes have only strings of letters and numbers that identify them in the massive catalogues they appear in. Millions of stars have been catalogued by photographic surveys with the world's largest telescope-camera systems.

Because the constellations' names form their genitives in a variety of ways, most amateur astronomers memorize the genitives, or at least a lot of them. A listing of the genitive for every constellation appears in Appendix 2. A list of the lowercase letters of the Greek alphabet is given in Table 1 below.

The following subsections of this Night describe briefly the brightest constellations of summer and autumn. Even if right now it is winter or spring, it is possible for you to observe some of these constellations. See Figures 17 and 18 for how (where and when) to do so. *Observe as many of the bright constellations of the seasons as you can, using Figures 17–18 as your visual guide. What features not described below do you notice?*

TABLE 1

The Greek Alphabet

α	Alpha	η	Eta	ν	Nu	τ	Tau
β	Beta	θ	Theta	ξ	Xi	υ	Upsilon
γ	Gamma	ι	Iota	o	Omicron	φ	Phi
δ	Delta	κ	Kappa	π	Pi	χ	Chi
ε	Epsilon	λ	Lambda	ρ	Rho	ψ	Psi
ζ	Zeta	μ	Mu	σ	Sigma	ω	Omega

The Summer
Triangle.
(Photographed
by Akira Fujii)

BRIGHTEST CONSTELLATIONS: SUMMER

Ascending ever so prominently up the east sky on summer evenings is the Summer Triangle of Vega, Altair, and Deneb. Vega's constellation is little *Lyra the Lyre* (a lyre is a kind of harp), and the bright star sits above a little diamond or parallelogram of modest stars. Deneb marks the tail of the graceful *Cygnus the Swan,* whose shape has also earned it the title "the Northern Cross." A very bright section of the Milky Way band wreathes much of the body of the Swan. Altair is in *Aquila the Eagle* and is flanked to either side by a moderately bright star.

Two bright constellations parade across the south sky on summer evenings. First is *Scorpius the Scorpion,* whose shape is something like a letter S fallen forward. Antares—flanked by two stars in a configuration resembling that of Altair—represents the slightly ruddy heart of this striking, though quite low, constellation. The sting of Scorpius is raised and tipped with two side-by-side bright stars, which some people have called "the Cat's Eyes." Following just behind Scorpius is *Sagittarius the Archer.* In really dark skies, the western part of Sagittarius is almost overwhelmed with bright patches of Milky Way glow, grandeur in the direction of our home galaxy's center. But the brightest stars of Sagittarius (including several of which are second magnitude) form a very obvious teapot pattern. We'll revisit this wondrous region when we talk about such things as the Milky Way and nebulae.

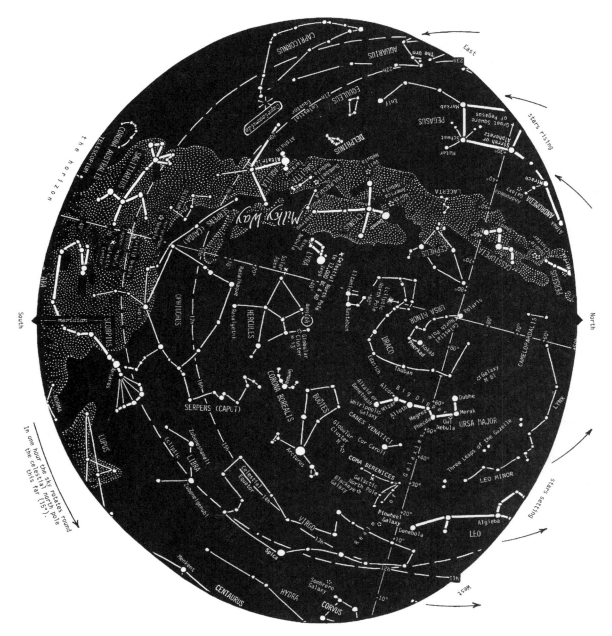

Figure 17.
Summer star map.
Map for July at
about 9 or 10 o'clock
in the evening.

BRIGHT CONSTELLATIONS: AUTUMN

The only truly bright constellation in the south sky on autumn evenings is *Pegasus the Flying Horse.* Pegasus has a second magnitude nose, the star Enif (Epsilon Pegasi), stuck way southwest from his body, but his body is his most famous part: its four second magnitude stars form "the Great Square of Pegasus." The Great Square is actually somewhat longer east-west than north-south. Extend a line down from the west side and it eventually leads to the star Fomalhaut, which is brilliant but in a feeble constellation.

The northeast (upper left) star of the Great Square belongs officially to *Andromeda the Chained Maiden.* This famous damsel in distress from Greek mythology features three equally spaced, almost equally bright second magnitude stars in a long line. It is also home to the naked-eye glow of the Great Andromeda Galaxy. The constellation Andromeda passes overhead at midevening in November for observers around 40° North. And her line of bright stars points to *Perseus the Champion,* who saved her from a monster in her legend. Perseus has a shape something like a letter K and contains two second magnitude stars, Mirfak (Alpha Persei) and Algol (Beta Persei). Algol is usually second magnitude. Actually, it is one of the few famous stars whose brightness varies. This eerie star rather appropriately marks the severed head of the monstrous Medusa, which Perseus is holding by its snaky hair.

The myth that includes Andromeda, Perseus, and Pegasus has several other representatives among the constellations. Brightest of the others is the zigzag or M shape we see high in the north on autumn evenings: *Cassiopeia the Queen,* mother of Andromeda. The compact pattern of Cassiopeia and its bright (several second magnitude, several almost that bright) stars is a circumpolar constellation high in the north when the Big Dipper is low.

When Cassiopeia and Andromeda are high, one leg of Perseus points down to the Pleiades, and Orion is rising in the east: winter and the whole year's cycle of bright constellations are soon to come again.

SUMMARY

Within each constellation, many of the naked-eye stars have been given scientific names that use a letter of the Greek alphabet followed by the Latin genitive form of the constellation's name (Alpha Orionis, Beta Tauri, etc.). The star designated alpha (first letter of the Greek alphabet) is often, but not always, the brightest star of the constellation.

The brightest constellations of summer are Lyra, Cygnus, Aquila, Scorpius, and Sagittarius. The brightest of autumn are Pegasus, Andromeda, Perseus, and Cassiopeia, which is really circumpolar.

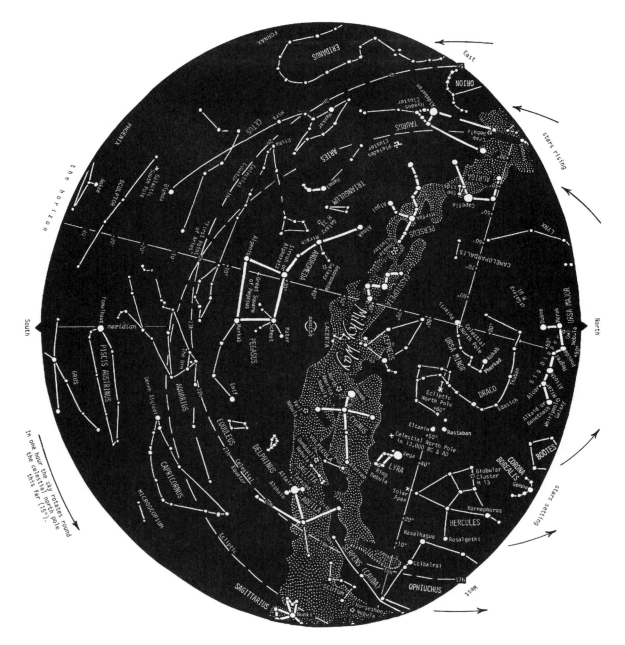

Figure 18.
Autumn star map.
Map for October at
about 9 or 10 o'clock
in the evening.

✦ NIGHT 17 ✦

DARKNESS OF THE SKY

✦ ✦ ✦

Time: Any time of night.

THERE ARE NIGHTS when clouds impede our view of the heavens. But many beginners might suppose that otherwise our ability to see the stars and other celestial objects would be similar from night to night and from place to place in the northern hemisphere.

In reality, part of the richness and variety of skywatching lies in how dark the sky appears tonight. Celestial objects shine, and in general the darker the sky background you see them against, the more prominently they will be visible.

But what causes the sky to be darker and therefore more star-filled on one cloud-free night than another, and at one observing site than another? Two things that can make the night sky less dark should spring to mind at once: moonlight and light pollution. You'll see only a small fraction as many stars when the Full Moon is in the sky, or where the glare from the poorly shielded lights of a big city brighten the atmosphere above and around you.

The third factor is less understood by most people. Astronomers call it the transparency of the atmosphere, and it is something you have to contend with even on a moonfree evening hundreds of miles from any artificial lighting.

Before you go out to judge how dark your sky is, let's discuss these three factors.

TRANSPARENCY AND ATMOSPHERIC EXTINCTION

Astronomers define **transparency** as the degree to which the atmosphere permits light to pass through it. The sky can be cloudless but still hazy. On a day of good transparency, the sky is dark and a deep blue; all we're seeing is the blue wavelengths of sunlight being scattered by air molecules. But on a day of bad transparency, the sky is bright, sometimes painfully so, and milky white in color (especially around the Sun). What we're seeing then is the combined scattering of all wavelengths of sunlight by humidity (water droplets) combined with particles of dust, pollen, or artificial pollutants. The night of good transparency contains stars, incredibly numerous and bright (if neither moonlight nor light pollution interfere). The night of bad transparency has a washed-out look, because much of the incoming starlight is being scattered; only a few stars are bright enough to be easily visible, and even those don't stand out well against the bright sky background.

Interestingly, what causes the problem is mostly scattering, not absorption, of starlight by haze. Or, at any rate, this is true if you are looking at a star or other celestial object at a fairly high angular altitude in the sky. At lower angles, especially down near the horizon, even a very transparent atmosphere ends up absorbing significant percentages of the star's light and dimming it greatly. This dimming is called **atmospheric extinction**. It is strong low in the sky because the light of a star low in the sky must travel through a longer section of atmosphere to reach us. Even on a quite transparent night, a star that is only 13° high in the sky shines only about half as bright as it does when it is 45° or higher! Atmospheric extinction is so strong within a few degrees of the horizon that only the brightest stars and planets can be glimpsed there with the naked eye.

Even the Sun is so greatly dimmed by atmospheric extinction that under normal conditions it ceases to be blindingly bright by the time it reaches the horizon.

MOONLIGHT

Most amateur astronomers want to know if the Moon will be up at a given time—in the hope that it won't be! This is shocking for the novice to hear. But the fact is that the Moon is often bright enough to make it hard to observe most of the dimmer lights in the sky. But don't carry this complaint too far and use it as an excuse to miss out on many a wonderful night! The crescent Moon a few days after or before new Moon has little effect on the brightness in most of the sky. Even when the Moon is half illuminated, most of the naked-eye sights you'd see on a moonless night can still be glimpsed as long as the atmosphere is quite transparent and you are looking at a point far away from the Moon's position. An interesting fact is that the Sun shining straight on the Moon when it is Full or near Full reflects back to us much more strongly than at smaller phases: the Moon is half as bright as Full not when it is half lit (seven days before full) but instead only two days before Full.

LIGHT POLLUTION

We discussed light pollution in Night 2. But there is more to say about it. First of all, consider the fact that the sky in a large city with typically wasteful lighting fixtures and practices is similar in brightness to that of a country sky with the Full Moon up. Remember that glare—light going directly in your eyes from local light sources—is worst of all, making it impossible for you to notice any but the brightest stars and planets.

It's alarming how far the effects of a city's light pollution can reach. Nevertheless, according to the basic light pollution formula called Walker's Law, your proximity to a city has a relatively greater effect than you'd think on the amount of light pollution your sky will suffer: getting twice as close to a city makes its skyglow appear about six times brighter.

Skyglow from cities typically lessens somewhat in the later hours of the night, when more businesses close and some lights do go off.

RATING YOUR SKY'S DARKNESS

Try rating the darkness of your sky tonight, and on other nights. For the most interesting results, estimate the sky darkness on a night when the stars are unusually numerous, a moonless night following a day of deep blue sky and low humidity. These are the optimum conditions for sky darkness at your location. To this base level, the variables of moonlight and haze would introduce additional brightness on many nights. What about light pollution? Do you find that your entire sky is any darker after midnight when some artificial light sources are turned off? Will next year's light pollution increase enough to cause a noticeable brightening of all or most of your sky? If you become keenly aware of how many stars and other night sky sights are visible at each of several levels of sky brightness, you can communicate to people what they will gain by each reduction of wasteful skyglow—and lose by each increase in it.

Before you try rating how faint the faintest stars you can see are, there are two key observational techniques to learn. The first is to allow for **dark adaptation**. You'll find that after ten minutes in a dark location you can see far fainter objects than when you first came outdoors, and after twenty minutes, even fainter ones. More time in darkness will increase your dark adaptation only a little more. So if possible do your rating after you've been out about twenty minutes. If you use a flashlight, cover it with red cellophane, because red light doesn't really reverse the chemistry of dark adaptation in your eye as bright light at other wavelengths does.

The second technique for seeing the faintest celestial objects you can is to use **averted vision**. Averted vision is looking just off to the side of a faint star or other celestial object, in order to let its light fall on the most sensitive parts of the eye's retina. It works!

I have produced a scale for rating the darkness of the night sky quickly. Here I will give a simplified version of it. We haven't yet discussed the Milky Way, but for our purposes here you only need to know that in summer and early autumn the brightest part of the Milky Way band is an arch of soft glow passing from approximately northeast to southwest high across the sky in the early evening. If your sky is dark enough, you will notice it easily. In addition, Figure 19 shows the magnitude of various stars in the Little Dipper, which is always partway up the north sky for viewers at midnorthern latitudes of Earth.

Which of the following describes your sky on a given night? Use the visibility of Milky Way, certain Little Dipper and other stars, and other features mentioned below to decide. Remember that even a "remote country sky" will look like a "city sky" on nights around Full Moon or when the sky is extremely hazy.

City sky. Many parts of the sky are extremely bright from light pollution and may even show a pinkish hue. Any clouds in the sky shine brilliantly. Only the three brightest stars of the Little Dipper are readily visible, if you shade your eyes from all local glare and know exactly where to look. No star, or at most one, is visible within the Great Square of Pegasus. There is no sign of the Milky Way, even in Cygnus when it is high in the sky.

Suburban (or small city) sky. At least a few parts of the sky are very bright from light pollution, but not high in the sky. Some clouds may be bright from light pollution; others are

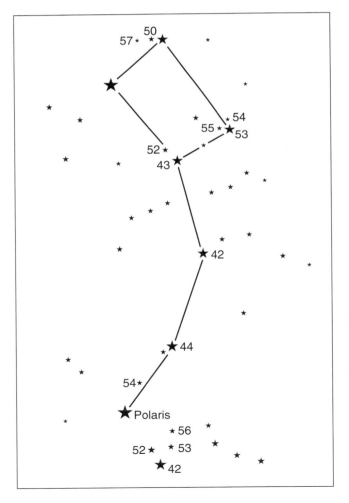

Figure 19.
Little Dipper magnitude scale.
Numbers are magnitudes
with decimal points removed
(example: "42" means a
magnitude of "4.2").

fairly dim. The fourth magnitude stars in the Little Dipper are readily visible, though dim, to the naked eye. A few stars may be glimpsed with some difficulty within the Great Square of Pegasus. A dim trace of the Milky Way is visible in Cygnus when it is high in the sky.

Country-near-small-city (or very small town) sky. There are some bright patches of light pollution, but only rather low in the sky. Most clouds glow fairly dimly from light pollution, about as bright as the Milky Way; occasionally clouds down low in the direction of cities are bright. The fourth magnitude stars of the Little Dipper are quite prominent and the fifth magnitude ones can be detected. Many stars are visible within the Great Square of Pegasus. Much of the Milky Way is visible, and its brightest parts are prominent.

Remote country sky. There are only a few fairly weak patches of light pollution glowing at spots along the horizon. Clouds fairly high in the sky glow dimly from light pollution, or they are dark. The fifth magnitude stars of the Little Dipper are easy to notice, and many fainter stars are detectable. The Great Square of Pegasus is swarming with stars. Faint, remote regions of the Milky Way band are visible, and much structure with many levels of brightness can be seen in the bright parts of the Milky Way. Many stars and the Milky Way are visible even fairly close to the horizon.

Conditions even better than those of "remote country sky" can be seen at the best observing sites or in parts of the sky of very good sites when the atmosphere is exceptionally transparent.

SUMMARY

The darker your sky is, the more stars and other objects you can see in it. The darkness of the night sky is diminished by haze, the atmosphere's greater thickness at a low angular altitude in the sky (strong atmospheric extinction), moonlight, and light pollution. The degree to which the atmosphere permits light to pass through it is its transparency. The intensity of light pollution at various distances from a city of any given population can be calculated with Walker's Law, which shows that distance is relatively more important than the strength of the light pollution source. The darkness of your sky on a given night and at a given site can be rated by diagnostic features such as the visibility of clouds, Milky Way, city skyglows, and stars of various magnitudes in familiar star patterns like the Little Dipper.

◆ NIGHT 18 ◆

THE ZODIAC CONSTELLATIONS

◆ ◆ ◆

Time: Any clear night.

IT WAS BACK on Night 8 when we first discussed the band of constellations in which we always find the Sun, Moon, and planets: the zodiac. Five of the twelve traditional constellations of the zodiac we have already examined individually as bright constellations. The other seven, which we'll consider on this current Night, are not so easy to observe if you live anywhere near a city and have only your unaided eyes to use. But they are interesting in many ways, and of course they are very famous because they are visited every month by the Moon, every year by the Sun, and at various intervals by the planets.

THE CIRCLE OF THE ANIMALS

First, let us consider the zodiac as a whole, as a series of constellations. According to Guy Ottewell, the word zodiac is from the Greek *zoidiakos kyklos,* "circle of little figures," from *zoidion,* "little animal," which comes from *zoon,* "living thing." The original Greek zodiac did consist entirely of animals, if we include humans as animals. In Roman times, huge Scorpius the Scorpion had his claws converted into Libra the Scales (presumably the scales of Roman justice); Libra remains the only inanimate figure in our modern zodiac.

As we learned in Night 15, the International Astronomical Union decided on the current official boundaries of the constellations in 1930. According to these boundaries, the ecliptic, the midline of the zodiac, passes through a thirteenth constellation, Ophiuchus the Serpent-Bearer, and very near to a corner of Cetus the Whale. As a matter of fact, the Sun spends a lot more time in Ophiuchus than in the neighboring traditional constellation of the zodiac, Scorpius. The amount of time the Sun spends within the modern boundaries of zodiac constellations varies greatly from one to another, with Virgo the Virgin hosting the longest stay.

Another change in our zodiac since its invention (or adaptation from still earlier cultures by the Greeks and Romans more than two thousand years ago) is the constellation that leads the zodiac. The first zodiac constellation is supposed to be the one that contains

TABLE 2

Constellations of the Zodiac

Aries the Ram	Libra the Scales
Taurus the Bull	Scorpius the Scorpion[a]
Gemini the Twins	Sagittarius the Archer
Cancer the Crab	Capricornus the Sea Goat (or Goat-Fish)
Leo the Lion	Aquarius the Water Carrier
Virgo the Virgin	Pisces the Fish

[a] The Sun actually spends less time in Scorpius than in the dimmer constellation Ophiuchus.

the vernal equinox: the position of the Sun at the start of spring (in the northern hemisphere). But precession, the extremely slow wobble of Earth's rotation axis, has altered that position from a spot in Aries to a spot in Pisces. Nevertheless, we still hear about "the First Point of Aries," and a modern verse that teaches the order of the zodiac constellations still gives the honor of leading to Aries:

> *The Ram, the Bull, the heavenly Twins,*
> *And next the Crab, the Lion shines,*
> *The Virgin and the Scales.*
> *The Scorpion, Archer, and the Goat,*
> *The Man who pours the Water out,*
> *And Fish with glittering tails.*

These constellations are Aries the Ram, Taurus the Bull, Gemini the Twins, Cancer the Crab, Leo the Lion, Virgo the Virgin, Libra the Scales, Scorpius the Scorpion, Sagittarius the Archer, Capricornus the Sea Goat (or Goat-Fish), Aquarius the Water-Carrier, and Pisces the Fish.

ZODIAC CONSTELLATIONS OF WINTER AND SPRING

Remember that the constellations we attribute to a given season are really just those that can be seen high in the evening sky during that season. If you want to see the constellations of the next season, all you have to do is go out later in the night.

There are a few constellations that stand on the border between seasons. Aries the Ram is often considered the last autumn constellation of the zodiac, but for our purposes we will take it to be the first of winter.

Go out and locate as many of the following constellations as you can, using the star maps of Figures 15–18. If you live in or near a sizable city, you may need the help of binoculars to trace these patterns, even on a clear, moonfree night.

Aries actually possesses several rather bright stars. The brightest is second magnitude Hamal, which a careful naked-eye look and any binocular look shows to have an orange hue. What makes Aries less than spectacular visually is the fact that its few brighter stars form a very small pattern, while the rest of the constellation is dim. Aries is often over-looked for the attractions of much bigger, brighter constellations in its general vicinity. Among these are Andromeda and Perseus and, somewhat farther off, Auriga and Orion. And it certainly suffers by comparison with the next two zodiac constellations, *Taurus* and *Gemini*, the former of which has a spectacular star cluster near Aries, the Pleiades, to draw attention away from the Ram.

Cancer is the next less-than-conspicuous zodiac constellation. It reaches the meridian in the evenings of early spring and the best word for it is "dim." It has sometimes been referred to as the void between Gemini and Leo, but this is only true if you are observing skies with fairly heavy light pollution. The center of the constellation is located almost halfway between the bright Gemini star Pollux and the bright Leo star Regulus. This center consists of several stars of average brightness and a star cluster, almost as bright, that can be seen in the country with the naked eye as a largish hazy patch of light if skies are clear. (We'll learn more about this M44 or Beehive Star Cluster later.)

After Cancer comes bright *Leo*, then after Leo, the long constellation *Virgo*. Virgo offers us the fine first magnitude star Spica. But its few other moderately bright stars are scattered over a large expanse of sky and not in any pattern that is easily formed or noticed. As we'll see later in this book, however, Virgo is a very interesting place to turn telescopes.

Libra is the next constellation after Virgo. Two of its moderately bright stars have common names that are among the longest and strangest, at least to English-speaking people: Zubenelgenubi and Zubeneschemali. Zubenelgenubi, more often known as Alpha Librae, consists of two component stars far enough apart for some people to glimpse both with the unaided eye. Zubeneschemali (Beta Librae) has a reputation of appearing slightly green-ish to some people—though you'll probably need binoculars at least to test whether your eyes detect green hue.

ZODIAC CONSTELLATIONS OF SUMMER AND AUTUMN

We could classify Libra as either a late spring or early summer constellation. The next two in the zodiac, *Scorpius* and *Sagittarius,* are certainly summer patterns and are both bright and glorious because of their many clusters and nebulae and their proximity to a bright part of the summer Milky Way. After them, however, comes a much dimmer constellation of either late summer or early autumn, *Capricornus*. The shape of Capricornus doesn't much resemble any familiar earthly object. Perhaps this encouraged stargazers very long before the Greeks to see the pattern as a strange amalgam of two creatures: front half a goat, back half a fish. The Sea Goat or Goat-Fish has some of its brightest stars bunched at either end of its pattern, whose shape perhaps most resembles a compressed boat. At the front (western) end is Alpha Capricorni (Algiedi), which like Alpha Librae can be split into two close-together stars by sharp naked eyes or any binoculars.

Following Capricornus is *Aquarius*. Most of this constellation's stars are also difficult to shape into a recognizable pattern—indeed some of them are supposed to represent streams of water being poured by the person, the Water Carrier. But at least his Water Jar or Urn is, though rather dim, a noticeable little pattern, shaped somewhat like a sideways letter Y. Unlike Capricornus, Aquarius holds some very interesting telescopic sights within its boundaries.

The final constellation of the zodiac is *Pisces*. Its stars are imagined to represent two fish, connected by a cord that holds both tails. Pisces is a very faint constellation, but with country skies or binoculars you should be able to find the roundish pattern of stars that represents the western fish's head; this pattern of stars is also an asterism called the Circlet. The Circlet shines right below the Great Square of Pegasus. The other fish points north toward Andromeda. And the knot in the cord that ties the two fish together is marked by the moderately bright star called Risha or Al Rischa.

SUMMARY

The constellations of the zodiac are, in order, Aries, Taurus, Gemini, Cancer, Leo, Virgo, Libra, Scorpius, Sagittarius, Capricornus, Aquarius, and Pisces. Aries contains the second magnitude orange star Hamal. Cancer is dim, is centered between Pollux and Regulus, and contains the Beehive Star Cluster. Virgo features first magnitude Spica but has few even moderately bright stars along its great length. Libra is composed of the stars Zubenelgenubi and Zubeneschemali, among others. Capricornus represents a creature that has the front half of a goat and the back half of a fish, and sharp eyes can see its alpha star as double in good conditions. Aquarius is mostly dim and sprawling but contains the noticeable asterism called the Water Jar or Urn. Pisces, two very dim fish, includes the noticeable asterism known as the Circlet, located right under (south of) the Great Square of Pegasus.

✦ NIGHT 19 ✦

A VARIETY OF NAKED-EYE PLANETS

✦ ✦ ✦

Time: Any time of night that a naked-eye planet is visible.

WE'VE DISCUSSED the motions and apparitions of superior planets in general and inferior planets in general. But even without a telescope to see their globes, the planets are all individuals.

Once you become a full-fledged amateur astronomer, you seldom forget where all the planets are, which constellations they are in, whether they are visible after sunset or before dawn, and which of them are especially bright and close right now. So it is not strictly necessary to learn how to identify one planet from another on the basis of appearance alone. But it is a tremendous pleasure to get to know each planet as an individual.

This Night you will simply observe the special characteristics of whatever planets are currently visible. The notes below tell you some of the properties to look for, depending on which planets you see and what they are doing.

THE BRIGHTEST PLANET

The first planet most people are likely to notice is Venus. Half the time—and for about nine months straight—it is visible only before dawn, when not many people are awake or outside. But then, for about nine months straight, Venus is glowing down near eye level from soon after sunset to as late as several hours after dusk. Even the most earthbound, self-absorbed person often notices it: a yellow-white point of light that outshines all others so obviously that there is no contest.

Venus is so bright that it sometimes casts shadows at remote country locations. It is so bright that you can often follow it with the naked eye until well after sunrise or find it well before sunset. When Venus is near its greatest brilliancy and your sky is deep blue, the planet can even be seen by the naked eye as a steely speck of light in the middle of the day.

The most dramatic changes in Venus's appearance, both in the telescope and with the naked eye, occur in the weeks just before and just after inferior conjunction. Most convenient to watch is the display before inferior conjunction, which occurs in the evening. One week Venus is as prominent as it has been for months, setting an hour and a half or two after the Sun; a week or two later, its brightness has dwindled and it is just glimpsed for a matter of minutes in bright twilight: what a dramatic exit from the evening sky! And a poignant one, because those of us who don't often get up before sunrise will be seeing far less of Venus for many months after that.

Wherever you see Venus in relation to the Sun and background stars, that's where it will be again almost exactly eight years later. This is due to the mathematical relation between the orbital periods of Earth and Venus. Thus, the very high, excellent evening apparition of Venus that occurred in the spring of 1996 will take place again in the spring of 2004.

THE DEPENDABLE GIANT

Jupiter is the second brightest planet, but whereas Venus can never be seen for more than a few hours before the Sun rises or after the Sun sets, Jupiter can often be viewed high in the middle of the night. Thus Jupiter is often the brightest point of light above the horizon.

Jupiter is most delightfully dependable in its telescopic appearance. It looms large with telescopic magnification, and even fairly small telescopes can reveal at least a belt or two of dark cloud, sometimes much more. But we are getting ahead of ourselves. Something reliable, and fascinating, about Jupiter as a naked-eye planet is that it advances almost exactly one zodiac constellation each year. This is because Jupiter takes almost twelve years to go around the Sun, and our heavens and the zodiac are divided into twelve parts, one for each month—though in one of these monthly parts it spends time in Scorpius *and* Ophiuchus.

SLOWEST OF THE BRIGHT PLANETS

Mention Saturn and many people immediately think of its gorgeous rings—which can't be seen with the naked eye. But you've got to find Saturn with the naked eye before you can turn your telescope on it. Saturn is a wonderfully unique naked-eye object in several ways.

Saturn usually ranges somewhere between 0.5 and 1.0 in magnitude, with favorable spells over the years when it becomes much brighter—a little brighter than zero magnitude. Add to Saturn's lack of twinkling its deep gold color and you have an object that seems decidedly somber or sedate—which is not the reaction you get when you see the telescopic view and those heart-stirring rings!

Another way in which Saturn seems sedate (or *saturnine* and gloomy!) was noticed by ancient cultures: this really is the slowest moving of the bright planets, the planets that were known in antiquity and before. Some Mesopotamian cultures called Saturn "the oldest of the old sheep" (the old sheep being the planets), because they had noticed its very slow movement among the stars. Saturn takes almost thirty years to complete an orbit. While Jupiter spends a tidy period of a year in each zodiac constellation, Saturn spends roughly two and a half years. Saturn spends most of each decade among the zodiac constellations of a single season. In the 1990s, Saturn has been a steady companion of the autumn stars. As we move into the twenty-first century, it will become a winter object.

As a matter of fact, the year 2000 is when Saturn's fellow giant planet, Jupiter, lumbers up to get in conjunction with it for the first time in about twenty years. Which reminds me of another "s" word that describes Saturn: solitary. Because Saturn is slower, it doesn't move into conjunction with the other naked-eye planets, and much of the year may go by with none of them near or moving into conjunction with it.

Even though we can't see the rings with the naked eye, we can notice changes in Saturn's brightness, some of which are caused by the rings being either more or less tilted toward Earth that season or year.

THE RED PLANET

Mars is the only planet with an almost ruddy hue. Call it the color of fire or of a pumpkin, there is certainly an orange tint to it. Unfortunately, when Mars is far from Earth, it's not bright enough for the naked eye to detect its color easily. At these times, you can use binoculars to note the color. Actually Mars is rather far from Earth for many, many months at a time, and during these times it may be confused with stars of similar brightness, the dimmest of the first magnitude stars. During these spells it moves rapidly through the constellations, one every month and a half or two, having conjunctions with many stars. And the payback for Mars's dim and distant times is the anticipation of its approach to Earth.

Mars is the planet that can vary its appearance the most over long periods of time. One reason is that it orbits the Sun at a speed not much slower than that of Earth. Thus it takes us a long time to catch up to it. But when we do, Mars can come closer than any planet except Venus. And unlike Venus, it isn't hidden in the glare of the Sun when it is closest to us; it is at opposition to the Sun and thus rises at sunset, is highest (culminates) in the middle of the night, and is visible all night long. Mars begins retrograde motion about a month and a half before opposition—sooner before opposition than any other planet—and makes a bigger retrograde loop than the other superior planets. In the months just before opposition, Mars brightens far more quickly than any other planet that can ever be seen high in the night sky. Each month, it almost doubles in brightness, its fiery hue becoming more prominent, its steady light glaring at us. It's no wonder that many ancient cultures associated Mars with their god of war; the impressive brightenings and seemingly blood-tinged hue must have been alarming indeed to people who couldn't explain them scientifically.

What made—and still makes—the brightening of Mars especially impressive is that at some oppositions it is much greater than at others. The synodic period of Mars is much longer than that of any other planet: Mars takes an average of about two years and two months to come back to the same position in relation to Earth and Sun. Thus one year tends to bring us an opposition and associated brilliant performance of Mars, the next a long period of modest brightness and less apparent color, then comes another opposition year. But why are some of Mars's oppositions so much brighter than others?

The explanation lies in the planet's orbit, which is much less circular than that of most other planets. The nearest point to the Sun on an orbit is called the **perihelion**; the farthest point from the Sun is called the **aphelion**. When Earth catches up to Mars for an opposition when Mars is near aphelion, the Red Planet brightens to a little less than the intensity of Sirius—a dramatic sight considering the planet's color and steady light. But after one of these so-called aphelic oppositions, each opposition of Mars keeps getting closer and brighter. Once every fifteen or seventeen years, Earth catches Mars when it is near perihelion. When this happens, Mars gets a great deal brighter than Sirius; it can even outshine Jupiter, to become the second brightest planet after Venus. The next of these perihelic oppositions occurs in 2003, when Mars comes a tiny bit closer to Earth than it has for thousands of years.

We'll return to the topic of close and far oppositions of Mars later in this book when we discuss telescopic observation and spacecraft exploration of this most earthlike and tantalizing of the other worlds. You don't have to have a telescope to look at Mars and thrill to the possibility of Martian life and of humans visiting Mars in a few decades.

THE ELUSIVE PLANET

One planet can shine brighter than Sirius. It is the one most often closest to Earth. Yet this planet has not been viewed by even some fairly devoted amateur astronomers.

Which is it? Mercury. Its relative closeness and brilliance do not necessarily guarantee that you'll see it easily because Mercury is so near the Sun in both space and the sky. The vast majority of the time Mercury is hidden in the solar glare. Mercury is usually only visible to the naked eye for about two weeks before and one week after an evening greatest elongation, and for about one week before and two weeks after a morning greatest elongation. In fact, Mercury may not be readily visible even in those periods unless the ecliptic makes a rather steep angle with the horizon at dusk or dawn at that time of year. For observers at Earth's midnorthern latitudes, the ecliptic is steepest in the dusk sky around the start of spring and in the dawn sky around the start of autumn.

It is a triumph just to glimpse this elusive, little, and fastest of all planets. Mercury rushes around its entire orbit in eighty-eight days. So there are a number of opportunities to see it either before sunrise or after sunset each year—but only a few good opportunities. Mercury sometimes takes on an orange color and quavers more than you'd expect a planet to. This is due to its location low in our sky, where it has to shine through a long pathway of the Earth's reddening and turbulent atmosphere.

SUMMARY

Each of the five bright planets known since ancient times presents unique, distinguishing sights to the naked-eye observer.

Venus is the brightest; it can often be seen with the naked eye long before sunset or long after sunrise. Its fall from the evening sky just before inferior conjunction, like its vault into the morning sky just after, is extremely swift and dramatic. Jupiter is the second brightest planet and the brightest that can be viewed high in a midnight sky. It moves one zodiac constellation per year. Saturn is much slower, taking most of a decade to cross the zodiac constellations of a single season. It shines sedately with a deep gold light, and its brightness is noticeably affected by the degree to which its rings are tilted toward Earth. Mars shines with a ruddier hue than any other planet. It alternates between one year of being distant and relatively inconspicuous and one year of brightening and fading dramatically from a very prominent display at opposition. At intervals of fifteen or seventeen years Mars has a perihelic opposition, in which it shines far brighter, becoming a truly imposing sight in the heavens. Mercury is the most elusive planet in our sky, pulling far enough away from the Sun to become visible for only a few weeks around its greatest elongations. The highest, most visible of these greatest elongations occur at dusk within a month or two of spring's start, and at dawn within a month or two of autumn's start.

✦ NIGHT 20 ✦

THE OTHER CONSTELLATIONS

✦   ✦   ✦

Time: Any time of night.

WE TOOK A LOOK at the brightest constellations on Nights 15 and 16, and the zodiac constellations on Night 18. Now it's time to survey some of the most interesting of the dozens of other constellations in the heavens. *Use the appropriate maps from Figures 15–18 to help you find the evening constellations of your current season—and the midnight or even dawn constellations of the next season or two if you feel enterprising! Compare what you see with the notes supplied below.*

Bear in mind that some of the constellations discussed here are quite faint. They look fine on a very clear, moonless night in a country sky, but near city lights or in less than excellent weather, a pair of binoculars will be required.

OTHER WINTER CONSTELLATIONS

Not just some but most of the traditional constellations of winter are bright. This is true of no other season. But several of the dimmer constellations of winter are particularly interesting.

Right under Orion, for instance, is *Lepus the Hare.* In the lore of the sky, this is naturally one of the prey animals that Orion or perhaps just his dog, Canis Major, is chasing. Lepus has several stars with proper names and is not really a faint constellation. If you have binoculars, you will see a star cluster and the distinctly ruddy R Leporis, "Hind's Crimson Star." Below Lepus is *Columba the Dove,* but a more important constellation lies to the west of Lepus. This is *Eridanus the River,* the longest of all constellations in the north-south dimension. It begins just west of Rigel and then meanders down, down, below the horizon of midnorthern latitudes, to finally reach its far-south first magnitude star Achernar.

Surrounded by such constellations as Orion, Canis Major, Canis Minor, and Gemini is *Monoceros the Unicorn.* There are no bright stars in the Unicorn, but a clear country night shows a wealth of fainter stars that are visible to the naked eye, and with binoculars and telescopes this is a real wonderland of star clusters and glowing clouds of interstellar gas.

Puppis the Poop (deck of the ship Argo) has some bright stars and fine star clusters, but is upstaged by Canis Major and is rather far south.

OTHER SPRING CONSTELLATIONS

Now we turn to dim constellations of spring. Among their number I include Boötes and Corona Borealis—though they might actually be considered bright.

Hydra the Sea Serpent, is the longest of all constellations in the west-to-east dimension. It represents a monster killed by Hercules as one of his great labors. When Hydra's head is well past the meridian, its tail is only just rising! Technically, Hydra covers more area in the sky than any other constellation. But very little of its vast expanse contains prominent stars or star patterns. The principal naked-eye attractions are the compact form of the Sea Serpent's head—which lies due south of Cancer—and the second magnitude orange star that marks the heart of Hydra: Alphard. After Alphard, the pattern of Hydra is faint and dives far under Corvus and Virgo to end under part of Libra. You might as well say Hydra stretches all the way from winter to summer.

Boötes the Herdsman has plenty of moderately bright stars—it's just that these are upstaged by the zero magnitude Arcturus. The main pattern of Boötes looks like a kite. Telescopes reveal many fine double stars in the constellation. One legend says that this herdsman is trying to protect his flock from the fierce Ursa Major. To help him, he has *Canes Venatici the Hunting Dogs,* a rather dim constellation mostly tucked under the tail of the Great Bear.

Between southern Boötes and Leo is *Coma Berenices, Berenice's Hair.* There are only rather dim stars here, but a number of them are gathered in a big, irregular star cluster that captivates the naked eye in very dark skies (otherwise, use binoculars). Coma Berenices is also famous for the many galaxies it offers for observers with telescopes. Its star cluster was once the tuft of Leo's tail but then became associated with the shorn locks of Queen Berenice, a real-life queen of Ptolemaic Egypt who sacrificed her hair to an altar in exchange for her husband's safe return from war. The story goes that the tresses disappeared from the altar and then were found twinkling in the sky as the Coma star cluster.

Well to the upper left of Boötes is a star pattern that used to be a sickle in the Herdsman's hand. It became *Corona Borealis, the Northern Crown,* a semicircle of stars adorned with one brighter gem, the second magnitude star Gemma or Alphecca. Corona Borealis is the crown of Ariadne, who in Greek mythology was abandoned by Theseus after she helped him escape from the Labyrinth. The story had a happy ending, for Ariadne was then courted by the god Bacchus, who saw to it that her crown eventually got to the heavens to honor her.

OTHER SUMMER CONSTELLATIONS

High overhead in early summer is *Hercules the Strongman*. His stars, only modestly bright, are spread over a large area, so he can be hard to find. Locate him about midway between the brilliant stars Arcturus and Vega. The most memorable part of the pattern of Hercules is "the Keystone" of four stars. Between the two on the west side lies a very special star cluster that can barely be glimpsed with the naked eye in dark skies but is dramatic in telescopes. South of Hercules is huge *Ophiuchus the Serpent-Bearer*, whom we have already encountered in connection with the zodiac. His brightest star is in his head, second magnitude Rasalhague. Ophiuchus separates the two parts of *Serpens the Serpent*, which like him contains some pretty sights for binoculars and telescope.

North of Hercules is one of the most underrated constellations in the heavens, *Draco the Dragon*. You'll have to turn northward to see this circumpolar constellation with its compact head facing Vega. The brightest star in this head is second magnitude Eltanin. The other stars in the head are third magnitude, fourth magnitude, and fifth magnitude. Draco contains a number of interesting objects for telescopes, but for the naked eye the major interest other than his head is in following his vast twisting coils. These form a kind of backward letter S with the last section of Draco thrust right between the Big Dipper and the Little Dipper. (This last section contains the star, Thuban, that was near the north celestial pole of the sky in the glory days of ancient Egypt.) Draco was once a larger constellation in several ways. Most remarkably, the Dragon once had wings: until about 500 B.C., they were our present-day Ursa Minor!

From Draco we fly far south to a loop of stars underneath the teapot of Sagittarius. This *Corona Australis the Southern Crown* is too far south to be seen well from the United States. Other little constellations of summer are much higher in the sky and are especially interesting for what can be seen in them with a telescope. *Scutum the Shield* is filled with a very bright patch of Milky Way and contains one of the finest star clusters that can be seen with a small telescope. *Vulpecula the Little Fox* is noted only for having the telescopic object known as the Dumbbell Nebula within its bounds. The other tiny, intricate constellations in or near the Summer Triangle are *Delphinus the Dolphin* (a minute diamond shape with a "tail") and *Sagitta the Arrow* (which does look like an arrow).

OTHER AUTUMN CONSTELLATIONS

Many of the autumn constellations are dim, and we covered most of them in our profile of the zodiac constellations. Just a few constellations of note remain.

We've mentioned *Piscis Austrinus the Southern Fish* as the constellation of the bright star Fomalhaut. It is otherwise nothing but a very dim loop of stars. Note that this Fish is a single individual, whereas the Fish of the constellation Pisces is plural (two fish).

High in the north in early autumn is the circumpolar constellation *Cepheus the King*. It is shaped like a little building, in autumn nearly upside-down when you face north, the sharp peak (steeple?) formed by a star only about twelve degrees from Polaris and the north celestial pole. Cepheus is a mostly dim companion for Cassiopeia but does contain second magnitude Alderamin and several very special stars that we will meet in upcoming Nights.

One final constellation based on a figure from the Perseus-Andromeda-Cassiopeia-Cepheus-Pegasus myth is *Cetus the Whale*. This constellation is usually identified as the sea monster that attacked Andromeda, which in mythology was not necessarily a whale. Cetus has nearly second magnitude Menkar in its head (formed by a fairly compact pattern of stars) and second magnitude Diphda (also called Deneb Kaitos, "tail of the whale") in its tail. But in between is a huge expanse that is sparsely starred. The neck of Cetus sometimes lights up with Mira, a remarkable star we'll learn more about later.

Tucked in between Andromeda and Aries is *Triangulum the Triangle*. Though small and dim, its triangle of stars is compact enough to be noticed in reasonably dark skies. And not too far from the point of the triangle is the interesting neighbor galaxy M33, a tricky but rewarding object for telescopes.

SUMMARY

Less bright but interesting constellations of winter include Lepus, Columba, Eridanus, Monoceros, and Puppis. Spring constellations of this sort include Hydra, Boötes, Canes Venatici, Coma Berenices, and Corona Borealis. Such summer constellations include Hercules, Ophiuchus, Serpens, Draco, Corona Australis, Scutum, Vulpecula, Delphinus, and Sagitta. Comparable autumn constellations include Piscis Austrinus, Cepheus, Cetus, and Triangulum.

✦ NIGHT 21 ✦

THE MILKY WAY

✦   ✦   ✦

Place: An observing site with no more than minor light pollution
(country-near-small-city level—see Night 17).
Time: Clear, moonless evenings, especially in summer and early
autumn, or, in spring, before dawn.

EVERYONE HAS HEARD of the Milky Way, and it has been impossible not to mention it a few times already in this book. But most people don't understand what it is. Also, in these days of widespread and severe light pollution, far fewer people have ever seen it. This is especially sad because it is one of astronomy's most beautiful and inspiring sights. This dreamy band of soft glow arching across the night is the immense structure of the magnificent galaxy we live in.

BAND AND GALAXY

The term Milky Way actually refers to two things. One is the band of glow that we observe in the heavens. The other is the **galaxy**—a vast aggregation of billions of stars revolving around a common gravitational center—in which we live. As seen from our vantage point, the band is the most prominent part of the galaxy. There are untold billions of galaxies scattered throughout the known universe, but the Milky Way is our home galaxy.

Even before we explore exactly how the band in the sky is related to the galaxy in space, we should observe the band. *At a fairly dark site (on a clear, moonless night), scan the sky for sections of a band of soft glow. Use the maps of Figures 15–18 as guides. On spring evenings, though, you may see no sign of it at all, even from a dark site. On the other hand, evenings in summer and early autumn present us with by far the brightest sections of Milky Way, which stretches high across the sky from the northeast to overhead to the southwest. Winter evenings offer a much fainter section of the Milky Way—but it's a delight to glimpse dimly wherever skies are very dark.*

The Milky Way band actually circles all the way around the celestial sphere. But at mid-northern latitudes on Earth the farthest south section—glorious in the Carina and the Crux (Southern Cross) and Centaurus regions—is forever hidden from us. What is the band?

A scan of bright sections of the Milky Way with binoculars in a clear dark sky reveals that the glow is formed by the combined radiance of innumerable stars too dim to be glimpsed individually with the naked eye. Research indicates that many of these stars are not just dozens or hundreds of light-years from us, the way most naked-eye stars are; they can be thousands of light-years away.

Why are so many of these distant stars confined to this band of sky?

The answer is that we live in a galaxy shaped like a lens, with its breadth much greater than its thickness (see Figure 20). When we look toward the Milky Way band in the sky, we are looking along the star-crowded equatorial plane of our lens-shaped galaxy, the plane which we are ourselves are in. On the other hand, when we look at right angles to this plane—to the regions of sky around Coma Berenices and Canes Venatici (galactic north)

Figure 20.
The Milky Way seen from side and face-on views showing approximate location of our solar system.

or below Cetus in Sculptor (galactic south)—we are looking up out of the star-rich plane and staring mostly into empty space beyond the Milky Way.

OUR SPIRAL GALAXY

We can say more about the form of our Milky Way galaxy. We have been able to piece together much of its large-scale structure and determine that it must be a spiral galaxy. We can see many spiral galaxies similar to ours when we look far off into space with telescopes. Spiral galaxies resemble pinwheels, with a central hub and spiral arms. The central hub is like a ball of billions of stars, mostly older, more sedate stars. The spiral arms are long, curving structures consisting of a wider variety of stars, including many spectacular big, bright, young ones, that continue to be born from the gas and dust—some starlit, some dark—in those arms. The entire structure spins around—about once every 200 million years or so at our sun's distance from the center. The Sun is believed to be about 26,000 light-years out from the center of the galaxy. Quite a few stars are as many as 50,000 light-years out from the center. Farther out, the stars become more sparse.

So we live maybe halfway out from the center of our galaxy as it might be observed by a viewer with a telescope located in some other galaxy. We live within a spiral arm, and our sky is more varied for it: there are impressive patches of superluminous stars in our sky (many of the stars of Orion are part of such a patch), and there are sections of the Milky Way band that are split or far more intricately structured by clouds of dark gas and dust. The most prominent split in the Milky Way band is caused by dark gas and dust, which divides the band into a main channel pouring south to Sagittarius and an apparent branch beginning in Cygnus that peters out in northeast Ophiuchus. This split is sometimes called "the Rift" or "the Great Rift."

There is too much light-absorbing gas and dust in the tens of thousands of light-years between us and the galactic hub for us to ever see it. But when we stare in the direction of the center of the galaxy, we at least see the glorious Milky Way star clouds of Sagittarius, which lie about 7,000 light-years inward from us. When we observe Orion, that fantastic region of superluminous stars and glowing gas, we look about 1,500 light-years into the middle of what may be an inward spur of the spiral arm on which we live. The "galactic center" is located in western Sagittarius. The "galactic anticenter" is located near the juncture of Taurus, Orion, Gemini, and Auriga. Thus, on summer evenings in Earth's northern hemisphere, we look inward toward the center of our galaxy, but on winter evenings we look outward, away from the center. The Milky Way band narrows in Perseus, but when we look out through our own spiral arm, we are staring at the Perseus arm, whose most famous resident may be the spectacular, bright Double Cluster in Perseus—about 7,000 or 8,000 light-years away.

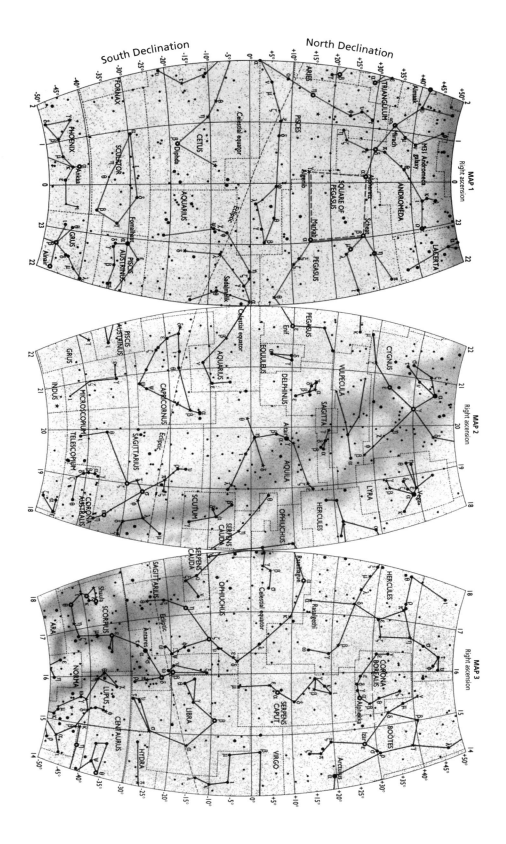

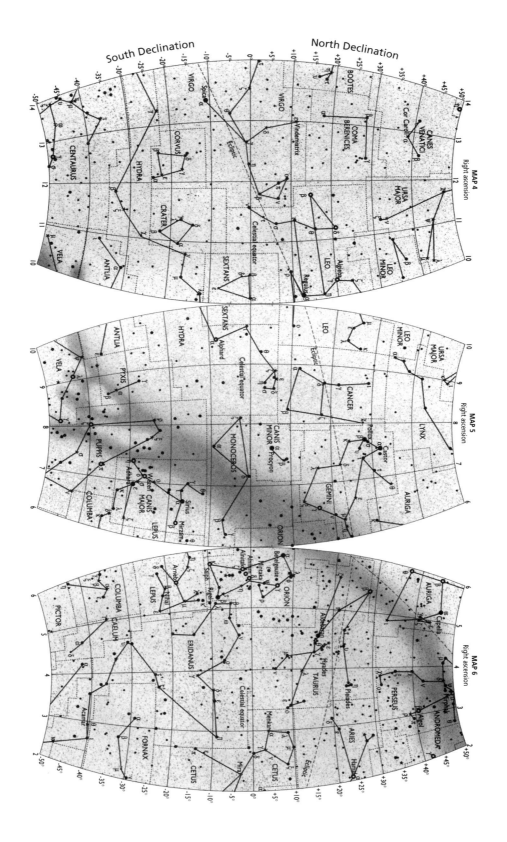

South Declination   North Declination

MAP 4
Right ascension

MAP 5
Right ascension

MAP 6
Right ascension

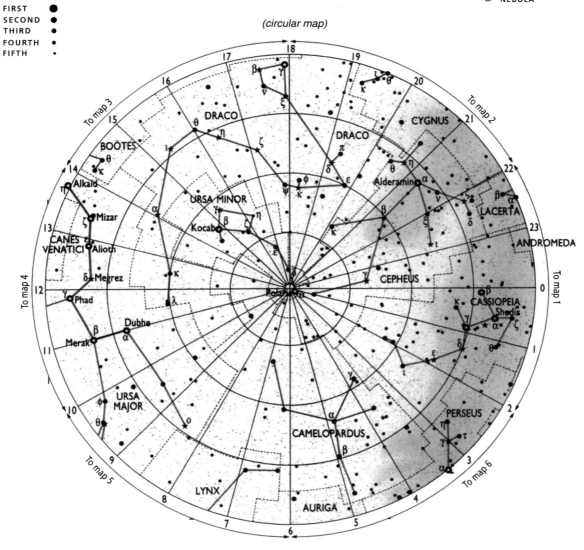

(circular map)

Figure 22.
North circumpolar region star map.

Bright section of
the Milky Way.
(Photographed
by Akira Fujii)

STAR CLOUDS OF THE MILKY WAY

The greatest concentrations of stars make up the brightest patches of Milky Way glow. These patches are called "star clouds." They are breathtaking to scan with binoculars or telescope. But they are awesome in a different way when seen with the naked eye.

Evening viewers will have to go out in summer and early autumn to observe star clouds. The one you're most likely to see—maybe the only you'll see with the naked eye if you live where there is considerable light pollution—is the Cygnus star cloud. The great advantage this cloud has over the others is that it passes overhead as seen from between 30° and 50° N, the most populous latitude band on Earth. Thus it can evade much of the effect of summer haze and light pollution found low in the sky. Even so, residents of very large cities cannot glimpse even the Cygnus star cloud amidst their awful light pollution.

About halfway down the south sky on July evenings is the roundish patch of the Scutum star cloud. The fact that it is quite compact and is bounded in part by dark clouds of dust and gas increases its visibility. If your view to lower in the south is blocked by buildings, trees, light pollution, or haze, the Scutum star cloud may impress you more than any other part of the Milky Way.

Oh, for a moonless night far from any cities after one of those infrequent cold fronts sweeps away the persistent haze of summer! Under these conditions, the two Sagittarius star clouds can be viewed in something approaching their proper splendor.

The large Sagittarius star cloud glows big just west of the spout of the teapot of Sagittarius. You may well mistake it for a cloud in Earth's atmosphere, peculiarly lit; the mind can hardly believe that this is a celestial phenomenon, a cloud composed not of countless water droplets but of countless stars. This cloud marks for us the direction to the hidden heart of the galaxy.

The small Sagittarius star cloud is much smaller: only about two degrees by one degree. But that's okay because this cloud's special glory is to appear as the densest, brightest knot of Milky Way. The small cloud is also called M24. (This designation used to be misapplied to an individual star cluster in the cloud, and you may find it as such in some older books.) Look for it not far above the top of the teapot of Sagittarius.

SUMMARY

The Milky Way is both the band of softly glowing light we see arching across the sky and the immense congregation of several hundred billion stars that is our home galaxy. The soft glow is composed of the combined light of countless distant stars a little too dim to glimpse individually with the naked eye. We see this glow in a band because we are looking along the star-rich equatorial plane of the galaxy that we ourselves are located in. The band of the Milky Way is split in places by regions of dark gas and dust. It is also spotted in several places with extra-bright patches of glow, which are star clouds. The Milky Way band is best seen on evenings in summer and early autumn.

✦ NIGHT 22 ✦

SPECIAL STARS

✦ ✦ ✦

Time: Any night.

STARS COME in various brightnesses and colors. The apparent brightness of a star depends upon how luminous the star really is and how far away. A star's color depends upon the surface temperature of the star; red stars are the coolest and blue-white stars the hottest.

But there are stars that are special not because they are unusually cool and red or because they shine 50,000 times brighter than our Sun. There are pairs of stars close together, called "double stars." There are other stars whose brightness changes dramatically, whether regularly or irregularly, called "variable stars."

Some of these special stars can be appreciated easily with the naked eye and form a good introduction to their kinds.

DOUBLE STARS

A **double star** is a point of light that upon closer or improved inspection turns out to be two or more stars very close together in the sky. They may also be close together in space, either orbiting each other or at least flying through space together. Or their apparent proximity may result because they lie on nearly the same line of sight from Earth, with one star much farther away.

A double star system in which the component suns are going through space together is called a **binary star**. A double star system in which there is merely a chance lineup in space, with one star often hundreds of light-years farther away than the other, is called an **optical double**.

There are other types of double stars, but these are more appropriately dealt with in the final section of this book. Two more terms that are vital in any discussion of double stars, however, are **primary** and **companion** (or **secondary**). The primary is the brighter of the two stars, not necessarily the more massive and gravitationally controlling.

Most double stars—especially binaries—are made up of components much too close together to "split" (separate into two distinct points of light) without optical aid. But there are a few naked-eye double stars in the heavens. The most notable is the Mizar-Alcor system.

Observe the Big Dipper—preferably when it is high and the sky is clear and dark—and examine the star at the bend in its handle. This magnitude 2.4 star, Mizar, is accompanied by a companion, magnitude 4.0 Alcor. Can your naked eyes detect Alcor, located just under 12 arc-minutes (about 0.20°) from Mizar? If you can't, maybe someone with slightly sharper vision can. And whether you can see Alcor or not tonight, you should try on other nights to see how sky conditions affect the observation.

There is a rich treasury of lore about Mizar and Alcor. They have been known as Horse and Rider. Another tale says that Alcor is the "lost" member of the Pleiades star cluster. A Norse account says that this is the toe of Orwandil (probably Orion), which Thor had to snap off when it got frostbite—and which he subsequently threw all the way to the north sky. For our purposes here, however, the most interesting lore about Mizar and Alcor is that one of the medieval Arabic names for the latter meant "the test." Was it a test of normally sharp vision or extremely acute vision? Probably average vision. In those days before the invention of eyeglasses, 20/20 vision would have been less common and more esteemed.

VARIABLE STARS

The stars were long regarded as the most changeless of all things. Lovers could swear by them, so enduring and constant were they. What do we now think? The star we call the Sun has shined for billions of years, and its energy output has not altered enough for the past few billion to endanger even the fragile thing we call life. But the brightness of many stars does fluctuate, in some cases enough for us to notice over the course of a year—in others enough for us to notice in an hour or less.

A **variable star** is a star whose brightness changes, whether regularly or irregularly. The amount of time between one peak or maximum brightness and the next is the **period**. The amount of change from minimum to maximum brightness is the **amplitude**. The line that shows brightness plotted against time on a diagram for a variable star is the **light curve**.

There are several different kinds of variable star, including those that suffer explosive loss of material in catastrophic outbursts. We'll meet with them later in this book. But for now, let's look for just a few of the brightest and most obvious variable stars that can be seen with unaided vision.

Two of the easiest to follow are Algol and Beta Lyrae. One or the other of them is above your horizon at some time on any evening of the year.

Algol is Beta Persei, the star that marks monster Medusa's severed head being carried by Perseus. In medieval Arabic the star's name means "the ghoul." Could it be that centuries

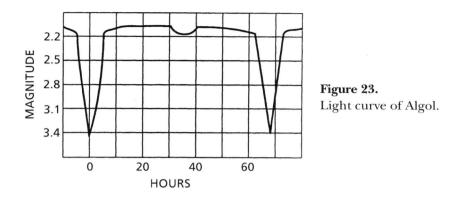

Figure 23.
Light curve of Algol.

or even millennia before the first recorded recognition of the star's variability (by Geminiano Montanari around 1667) the behavior of this star was noticed and found eerie and alarming?

Algol usually appears as a magnitude 2.1 star, just a bit dimmer than Mirfak (Alpha Persei). But about every third night (the period is about 2 days, 20 hours, 48 minutes, and 56 seconds) the star can be found to have dimmed down to magnitude 3.4 after a five-hour-long fade (followed by a five-hour-long brightening back to 2.1). This is a dramatic change, but you'd be surprised at how relatively seldom you'll catch it if you don't look night after night—or, more easily, look up the predicted dates and times of "minima" of Algol in the "Calendar Notes" section of *Sky & Telescope* magazine (see Sources of Information). For Algol's typical light curve, see Figure 23.

Beta Lyrae is only as bright at maximum as Algol is at minimum, but its variations can be easily detected because there is a convenient nearby **comparison star**, a star of constant brightness with which to compare the variable star. Beta Lyrae and its comparison star, Gamma Lyrae, form the southern end (end farthest from Vega) of the little diamond pat-

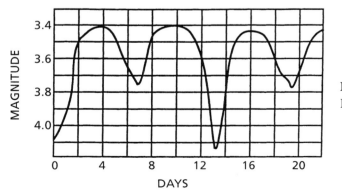

Figure 24.
Light curve of Beta Lyrae.

tern of Lyra the Lyre. Beta Lyrae (which is, very occasionally, called Shelyak) is closer to Vega than Gamma Lyrae (very occasionally called Sulafat).

Gamma Lyrae shines at a constant magnitude 3.25; Beta at maximum is 3.4, almost identical in brightness with Gamma (the trained human eye can only distinguish about 0.1 magnitude difference between two objects). About every thirteen days, Beta Lyrae reaches the minimum of a fairly swift fading and shines at magnitude 4.1—far dimmer than Gamma. But this is not Beta's only minimum. About halfway between its deeper dimmings, it falls to a minimum of magnitude 3.8—not drastically but quite noticeably dimmer than Gamma. (Figure 24 shows a typical light curve of Beta Lyrae.)

Go out and find Algol on a night when you know it is predicted to be dimming or brightening. If you can't get this information, check the star out to see if you can get a little lucky and catch it in the process of varying. Does it look dramatically dimmer than Alpha Persei? If Algol is not above the horizon this evening, Beta Lyrae will be and you can check it. And Delta Cephei is another important variable star you can check (see light curve in Figure 25). It is at a different brightness every night and, being circumpolar, is always visible to most of the world's population.

What causes the brightness changes of variable stars? In some cases, what we are seeing is a binary star system in which one of the suns is periodically eclipsing the other or each eclipses the other. Algol is such a star; Beta Lyrae is not as simple and is thought to be composed of two highly ellipsoidal stars so close together that a streamer of material from one is being sucked into the other. In other cases, a star is undergoing actual pulsations, changing its true size and brightness. Delta Cephei is the prototypical example of one class of such stars. But there are several major mechanisms, and many types of variable star. We'll explore some of these later in the book.

SUMMARY

A double star looks like a single point of light until closer inspection shows it to be a pair of stars, usually very close together in the sky. If a double star consists of a pair of suns actu-

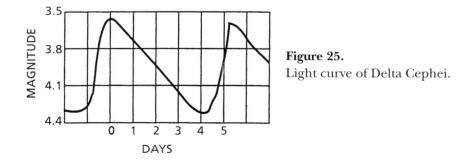

Figure 25.
Light curve of Delta Cephei.

ally traveling through space together (maybe even orbiting around each other), it is called a binary. If a double star consists of two stars that just happen to lie along the same line of sight but are actually at vastly different distances from Earth, it is called an optical double. The brighter star in a double star system is the primary, the dimmer is the secondary or companion.

A variable star varies in brightness. The time from one maximum of brightness to the next is the period; the amount of change from minimum to maximum is the amplitude. The line representing the star's changing magnitude plotted against time is the star's light curve. There are many kinds of variable star, including those whose brightness changes due to physical pulsations and those whose brightness changes due to eclipses between the components of a binary star system.

+ NIGHT 23 +

THE BRIGHTEST DEEP-SKY OBJECTS

+ + +

Time: Any night, preferably a clear,
moonless one far from city lights.

WHAT IS A "deep-sky object"? You'll hear advanced amateur astronomers use this term a lot. Its most liberal definition is "any object beyond our solar system." In practice, however, individual stars or star systems are usually excluded—no variable stars or double stars. If the term "deep-sky object" was not invented by the late and great *Sky & Telescope* columnist Walter Scott Houston, it was certainly made popular by his marvelous "Deep-Sky Wonders" column. And that column rarely mentioned double stars or variable stars, even though the observation of variable stars was a tremendous passion of his.

What else is outside our solar system besides stars? The answer: star clusters, nebulae, galaxies—and all three occur in different varieties.

We've already discussed what a galaxy is. But what is a star cluster, and what is a nebula (plural: "nebulae")? We'll use this Night to find out. Most deep-sky objects require binoculars or telescopes to see or at least to see well. But this Night we'll meet with those especially prominent ones that do not.

THE MESSIER OBJECTS

There is an important subset of the "deep-sky objects" class that we should discuss first: the Messier objects. No, these are not the objects more tattered, disheveled, or sloppily arranged than others. The Messier objects are those catalogued by the eighteenth-century French astronomer Charles Messier (1730–1817). English-speaking astronomers pronounce the second name as the French would: MESS-ee-ay.

Messier's primary interest was discovering new comets. When first detected, comets often do not display a tail but rather appear only as fuzzy patches of light. Unfortunately, at the low magnifications that are best to use when hunting for comets, most deep-sky objects also appear as hazy smudges of light and are easily mistaken for comets. This is precisely what happened to Messier: he thought he had found a comet; only after some time passed and it had failed to move in relation to the background stars did he realize it was a deep-sky object. Charles Messier started his catalogue of deep-sky objects principally to keep from mistaking them for comets. And yet today he is best known for having provided amateur astronomers with a handy list of many of the finest star clusters, nebulae, and galaxies that can be easily seen in small telescopes.

Many of the objects in Messier's catalogue were not first discovered by him. And in recent decades there has been some dissension about adding a last group of deep-sky objects that Messier saw to the earlier, shorter version of his list. Nevertheless, today most amateur astronomers agree that the longer list of 110 "Messier objects" or "M-objects" makes a great starting place for observers with small telescopes who wish to begin some scenic touring of the universe beyond our solar system.

I keep mentioning telescopes. M1 (the first object in the Messier catalogue) is a nebula (technically speaking, a supernova remnant) that can only be seen in a telescope. You'd also best have optical aid if you want a proper look at M2, M3, M4, and the vast majority of other M-objects. But there are, as we'll see, some glorious exceptions.

OPEN AND GLOBULAR STAR CLUSTERS

A **star cluster** is a group of stars traveling through space together. The gravitational bond among the members is strong enough for most to remain in one another's vicinity for hundreds of millions or billions of years after their birth.

There are two major kinds of star cluster. An **open cluster** (also called **galactic cluster**) is a relatively loose ("open"), more or less irregularly shaped arrangement of stars, typically between a few dozen and a few hundred. A **globular cluster** is a much larger, more tightly packed and spherical gathering of stars, typically containing somewhere from tens of thousands up to as many as several million stars.

Globular clusters would be staggeringly brighter and more glorious to the naked eye and to telescopes than open clusters if not for the fact that there are far fewer globulars and even the closest are at great distances from us. The closest open clusters are a few hundred light-years away; the closest globulars are about 10,000 to 20,000 light-years away. Consequently, several open clusters are among the sky's most prominent asterisms, visible even from cities, whereas no globular cluster that can be seen from 40° N is visible at all to the naked eye except from a rather dark country site.

NAKED-EYE STAR CLUSTERS

Other differences between open and globular star clusters are extremely interesting and are discussed later in this book. For now, *try to observe as many as of these bright naked-eye star clusters as you can, depending on the time of year.*

The Pleiades (M45). Named object number 45 in Messier's catalogue, but known from before the beginning of history, this loveliest of all naked-eye star clusters is the most famous. It has the finest combination of brightness and concentration of any naked-eye gathering of stars, making it a unique sight. Cultures have used it as a basis for starting their year. The Druids did so, and Halloween occurs when it does because this was when the Pleiades rose at nightfall and were highest at midnight. The cluster appears in everything from the Bible and Homer down to the stylized design of them on Subaru cars— "subaru" is the Japanese word for the Pleiades. Is your unaided vision sharp enough to distinguish the two close-together stars that form the tiny handle of the cluster's tiny dipper shape? Under good sky conditions, most people can count six or seven individual stars in the cluster with the naked eye, but more can be glimpsed if conditions are excellent and you know exactly where to look. The main part of the cluster is about one and a half degrees across. (Autumn to early spring.)

The Hyades. This cluster is even closer to us than the Pleiades. Although its total apparent magnitude is greater, it is not as concentrated and so looks a little less impressive. The Hyades are striking in a different way. They outline the handsome face of Taurus the Bull and one arm of the V-shaped pattern of the cluster's brightest stars ends with first magnitude Aldebaran, which is not really a member of the cluster but adds to its splendor. The main stars of the Hyades are gathered within about seven degrees of each other. How remarkable that the Hyades and the Pleiades—the two finest clusters close to Earth (about 150 and 400 light-years away, respectively)—should shine close together in our sky in Taurus. (Autumn to early spring.)

The Beehive Cluster (M44). In good binoculars or a small telescope, the open cluster M44 looks like a swarm of glittering bees. It is known as both the Beehive and Praesepe. Praesepe, the far older title, means the "manger"—for the stars Asellus Borealis and Asellus Australis, two asses or donkeys, to come feed at. To the naked eye, M44 looks like a large hazy patch of light, the many individual stars almost completely blending together. The cluster lies in the midst of Cancer the Crab, very close to the ecliptic and so a target for planets. (Spring, early summer.)

Coma Star Cluster. This cluster is almost as big as the Hyades, but the naked eye needs quite dark country skies to see it properly. For the legend behind this open cluster and its constellation, Coma Berenices, see Night 20. (Spring, summer.)

M13. This is the globular cluster that is easiest to glimpse with the naked eye from the latitudes of the United States—it passes very high. You'll see just a hazy glimmer of light about two-thirds of the way from Zeta to Eta Herculis (the line that forms the west side of the Keystone asterism of Hercules). M13's magnitude is usually listed as 5.7, so you'll need quite dark skies and averted vision to glimpse it. (Late spring, summer.)

M7. Although low in the sky, about four degrees northeast of Lambda Scorpii in the Scorpion's tail, this big third magnitude open cluster is easily visible to the naked eye as a sizable fuzzy spot of glow on clear, moonfree evenings. (Summer.)

Double Cluster in Perseus. Although best known as a spectacle for telescopes, this pair of clusters is also an intriguing sight to the naked eye in reasonably dark skies. Look for a slightly elongated, almost barbell-shaped patch of glow about midway between the main patterns of Cassiopeia and Perseus. (Autumn, winter.)

Alpha Persei Cluster. The brightness of second magnitude Alpha Persei (popularly known as Mirfak) almost overwhelms the other stars of this large and bright gathering, but it is a fine sight in dark skies. (Autumn, winter.)

NEBULAE AND GALAXIES

We defined "galaxy" when we discussed the Milky Way in Night 21. A **nebula** is a cloud of gas and dust in interstellar space. Few nebulae and galaxies are bright enough to glimpse with the naked eye, so these will be discussed further the in the section of this book concerning sights that can be seen by telescope. For now, *look for the following nebulae and galaxies with your naked eye in dark country skies.*

The Great Nebula in Orion (M42). We've already mentioned this one. You'll be thrilled if you can make out its hazy glow burning around the middle star in the Sword of Orion. But this amazing birthing place of new stars really needs to be seen in telescopes, which reveal it as probably the most complex and grand of all deep-sky objects. (Winter.)

The Lagoon Nebula (M8). Though not as bright as M42, M8 also has less interference to naked-eye visibility from nearby stars. On any night that you can see the large and small

The Great Nebula
in Orion. (USNO
photograph)

Sagittarius star clouds, you'll also see M8—like a little puff of smoke—coming up from the "spout" of the teapot pattern of Sagittarius. (Summer.)

The Great Galaxy in Andromeda (M31). This is the big sister of our Milky Way galaxy. Roughly twice as large as ours, the Andromeda galaxy looks like an elongated smear of glow that may be as long as four or five degrees to the naked eye in very dark skies. M31 lies almost three million light-years from Earth—the light you see from it tonight started out from M31 almost three million years ago! (Autumn.)

SUMMARY

Deep-sky objects are objects beyond our solar system, though in practice stars are usually excluded. Messier objects (also called M-objects and designated M1, M2, and so on) are a few more than one hundred especially bright deep-sky objects catalogued by eighteenth-century French astronomer Charles Messier. A star cluster is a group of more than just a few stars gravitationally bound to each other closely enough to stay together for up to billions of years before drifting apart or dying out. Open (or galactic) clusters are looser arrangements of dozens to hundreds of stars, but they are much closer to us and thus generally appear brighter than the huge globular clusters, which are tighter, more spherical arrangements of tens of thousands or even several million stars. Nebulae are clouds of gas and dust in interstellar space.

✦ NIGHT 24 ✦

METEORS

✦   ✦   ✦

Time: Any time of night for "sporadic" meteors.

WE'VE ALREADY encountered meteors briefly in this book. And, indeed, in real life we do encounter them briefly—usually for no more than a fraction of a second. Yet that fraction of a second may contain a glittering, flaring, vividly colored mass of incandescence hurtling across the heavens and vanishing, leaving a glowing trail of wonder that lingers in the sky for an incredible full second or two—and in the mind for years to come.

Meteors are the fireworks of the heavens. Unfortunately, many people have so seldom looked into a starry sky that they have never seen one of these so-called shooting stars or falling stars. Nor do most people realize that there are nights—the same nights every year—when the chance of seeing meteors increases tremendously. These are the nights of the meteor showers, and they are a sort of holiday in the astronomer's year. And yet patient watching on any clear dark night may also reveal more meteors—meteors that are not part of showers—than the beginner would have believed possible.

WHAT IS A METEOR?

We know that a meteor is not a star falling from the heavens. But what exactly produces the startling phenomenon we experience?

A **meteor** is the shining streak (or similar light) produced when a particle of silicate material and/or iron enters Earth's atmosphere at stupendous speed and burns up from friction with the air. (This is a simplified definition but will serve us well enough.) Most meteors are luminous at altitudes of fifty miles or more, and rarely survive much deeper into the atmosphere. Even a meteor the size of a grain of sand is capable of producing a streak as bright as a first magnitude star as seen by a well-placed observer below. But the explanation lies in their velocity: meteors that encounter Earth head-on or nearly head-on enter the atmosphere at speeds well over 100,000 miles per hour (no wonder they burn bright and don't last long!). Meteors sometimes leave a lingering trail that has a soft, eerie glow. This is caused by their passage having ionized (removed or added electrons to give atoms an electric charge) atmospheric gases. Such a trail very rarely lasts for more than a second, or a couple of seconds. It is called a **meteor train**.

When the particle is still out in space, it is called a **meteoroid**. It may be as small as a grain of dust or as big as a football field. Some meteoroids are pieces from the larger rocky bodies called asteroids, and many—especially those in meteor showers—are the larger specks of dust released by those gas-and-dust-spouting mountains of vaporized ice we call comets. (Most comets and asteroids are too dim to see with the naked eye millions of miles away in space, but there are some amazing exceptions we'll consider later in this book.)

A **meteorite** is the name given to the object if it survives its passage through the atmosphere to reach the surface of the Earth—an exceedingly rare event. Meteorites usually slow to the point where they cease shining and then fall the rest of the way dark and cold.

Various names are given to different kinds of meteor. A meteor brighter than any of the stars or planets (thus brighter than −4, the magnitude of Venus) is called a **fireball**. The term **bolide** is sometimes applied to a meteor that explodes or bursts. But the most interesting distinction between meteors is between those that come in showers and those that come sporadically, which appear in isolation. We discuss the differences below, but this Night is devoted to observing just the sporadics, including those that are spectacularly bright. In our next Night, we will observe the specific meteor showers.

METEOR SHOWERS AND SPORADIC METEORS

A **meteor shower** is a large number of meteors all seeming to radiate from a specific point or region among the constellations, called the **radiant**. Meteors that belong to the shower usually do not become visible at the radiant itself but rather at some distance from it, yet if you traced an imaginary line back from where they became luminous, it would eventually take you back to the radiant. Why do meteors in a shower shoot out from a particular spot? It is actually an effect of perspective. The meteors really are coming more or less parallel to each other, but they appear to radiate from a point, like snowflakes seem to originate from one point in front of the windshield as you drive through a snowstorm. Perhaps an even better example is supplied by railroad tracks. We know that the rails are parallel, yet perspective makes them appear to diverge from a point in the distance, as seen in Figure 26.

Each year offers its annual meteor showers on approximately the same dates. What causes this timely production of a meteor shower—or, indeed, meteor showers in general? A **meteoroid stream** is a very roughly cylindrical bundle of meteoroid paths in space. It is typically produced by the diffusion of particles from a comet's orbit. The particles may be scattered fairly regularly all along the orbit, but in some cases they form especially dense **meteoroid swarms** here and there—most often within a year ahead of or behind the comet itself. Some of the best meteor showers can produce dozens of visible meteors per hour

Figure 26.
Radiant of meteor shower and railroad tracks. Like the rails, meteors from a shower appear to diverge from a point in the distance even though they are traveling parallel paths.

when Earth crosses through them each year on the same dates. But if Earth encounters a relatively dense meteoroid swarm, observers might see up to thousands of meteors per hour. This very rare spectacle is called a **meteor storm.**

On the other hand, every night brings some meteors that come from no known radiant. Some of these are probably left over from a long-dead comet's meteoroid stream, which has now diffused widely across space. A meteor that cannot be traced to any known meteor shower is called a **sporadic meteor.**

HOW TO OBSERVE AND COUNT METEORS

There are plenty of useful tips to know if you wish to see more and better meteors. For instance, the best time of night to see meteors is between midnight and dawn. Why? Because Earth's rotational motion is then added to its orbital motion, so that meteors encountering us head-on enter our atmosphere at greater speed and therefore tend to be brighter and more numerous. But a few of the longest-lasting meteors are the slow ones just after nightfall. By going slowly the larger among these are more likely to survive to the lower atmosphere and there blaze at tremendous brightness. Notice also that most of the major meteor showers (see Table 3) have their radiants highest in the hours before dawn. Where should you look in the sky during a meteor shower? Near the radiant, meteors may be concentrated into a smaller area, but their luminous paths are shorter; far from the radiant, luminous paths are longer but the meteors are spread too far apart for easy observation. A moderate angular distance (say thirty to forty degrees) from the radiant is the best compromise.

Using a lawn chair, groundcloth and sleeping bag, or similar arrangement, go out on a clear and moonless night, preferably with limited light pollution, when no meteor shower is due. During an hour's patient perusing of the heavens, you may see just one or two but possibly five or even ten sporadic meteors.

You can, of course, just enjoy the thrill of the dark night being broken—repeatedly!—by the gliding streak of the meteors. Or you can also keep careful count of how many you see. Each person watching should keep a separate tally of his or her meteors—not counting any that would not have been seen if not for another person's sudden alerting cry. Be sure also to keep a separate score of sporadic meteors and shower members. Try to observe for regular intervals of time, an hour, half hour, or quarter hour. Do note when and for how long you take breaks.

SUMMARY

Meteors, popularly known as shooting stars or falling stars, are the streaks of light we see when a particle of space rock or iron enters Earth's atmosphere at tremendous speed and burns up from friction. When still in space, the object is called a meteoroid; in the rare cases it reaches the surface of the planet, it is called a meteorite. A meteor shower is a large number of meteors that seem to come from a single point or area, a radiant, among the constellations on the same dates each year. Meteor showers occur when the Earth crosses through the meteoroid stream that has diffused from the orbit of whichever comet originally released the meteoroids as its ice vaporized. When denser concentrations of meteoroids exist along a meteoroid stream, they are called meteoroid swarms. When Earth encounters a meteoroid swarm, observers may see a meteor storm featuring thousands of meteors per hour. Meteors that do not belong to any known shower are called sporadics; patient observing in dark skies can reveal ten or more of them an hour. Each observer should keep a separate tally, not counting meteors that wouldn't have been seen without a companion's alerting cry. An observer should try to make counts in orderly fractions of an hour, like fifteen minutes, thirty minutes, or one hour.

✦ NIGHT 25 ✦

METEOR SHOWERS

✦   ✦   ✦

Time: Certain hours of certain nights each year (see Table 3).

IF YOU'RE SUPPOSED to get a wish whenever you see a falling star, imagine having dozens of wishes per hour at your disposal. Such is the case when you are treated to a fine display of one of the year's best meteor showers.

SPECIAL POINTS FOR SHOWER WATCHERS

In the previous Night, we covered a lot of information about meteors in general and about how to observe and count them. But there are some additional points that are important specifically for observing meteor showers.

As you look at Table 3 and study when and where showers are at their best, you'll notice a few interesting trends. Observe that the radiants of most of the major meteor showers are highest in the hours before dawn. This is partly due to a fact we discussed in the previous Night, that there are more meteors to be seen toward dawn because Earth's rotational speed is added to its orbital velocity. It's much harder for a meteor shower coming at Earth from the evening direction to produce enough visible meteors to be major. The Taurid shower doesn't offer us very high rates, but it does seem to offer a more than usual number of fireballs. This is probably just a consequence of the size of its particles, depending on the nature of its source, which appears to be Comet Encke. The Geminid meteor shower—often the year's best—has its radiant highest earlier in the night than most showers. And the situation is even better than it looks. For midnorthern latitudes, the radiant in Gemini is high enough by midevening to make visible most of the meteors from it.

But now notice two other dramatic imbalances in the table: there are far more major showers from the northern celestial hemisphere than the southern, far more major showers in the second half of the year than the first. Why? No one has been able to come up with a persuasive explanation yet. But the seasonal anomaly reminds us of the poet Milton's great lines:

Swift as a shooting star
In Autumn thwarts the night.

So you are going to be doing a lot of your shower watching in the latter part of the year, very late at night or very early in the morning. (Meteor watching often means dressing your warmest, and bringing a friend for company!) But where should you be looking to see shower meteors? You can look almost anywhere in the sky and see some of them. But where do you usually maximize your chances of seeing the greatest number? As explained in Night 24, a moderate angular distance (say thirty to forty degrees) from the radiant is the best compromise.

Having said this, though, I will now turn around and argue that on some shower nights you may wish to look far from and close to the radiant, simply for the marvelous variety. The long, lazier flights of the meteors on the opposite side of the sky from the radiant are fascinating. An even stronger word is needed to describe the experience of seeing meteors very near the radiant. What you get are point meteors: a point or flash of light appears and may linger a bit longer than you think it should. What you have seen is a meteor that was heading directly toward you!

The night of predicted maximum can vary slightly, both in reality and for purely calendrical reasons like leap year, so we should try to observe the night before and after these dates. In fact, some shower members may be seen many nights or weeks before the peak night, and a good measure of when a shower will be interesting is the period it is predicted to be above quarter strength (one-quarter as many meteors as at maximum).

THE SHOWERS THEMSELVES

Go out on the night and at the time that a meteor shower is expected to be at its best. Count the meteors you see coming from the radiant. Keep a separate count of those from other showers' radiants and those that appear to be sporadic.

The following notes about the individual showers will assist you. Refer to Table 3 for basic statistics about each shower.

Quadrantids. These meteors come from a point in the long-abandoned constellation Quandrans Muralis (the Wall Quadrant), a point now located in northern Boötes. The shower is remarkable for being very strong but with very brief peak activity. Rates of 50, 100, even 150 Quadrantids per hour have been observed. But this only occurs for a few hours, so it is only seen in years when the brief peak happens to fall just before dawn when the radiant is high.

Lyrids. This shower is the earliest recorded in history, having been noted by the Chinese in 687 B.C. In modern times, the Lyrids seem to be weaker than they once were, but there

TABLE 3

Selected Meteor Showers

Shower	Maximum	Above One-Quarter Maximum[a]	Some Visible	Number per Hour[b]	Time[c]	Radiant[d]
1. Quadrantids	Jan 4	Jan 4	Jan 1-6	40	6:00 A.M.	15h28m, +50°
2. Lyrids	Apr 22	Apr 21-23	Apr 18-25	15	12:00 A.M.	18h4m, +34°
3. Eta Aquarids	May 5	May 1-10	Apr 21-May 12	10	4:00 A.M.	22h30m, -2°
4. Delta Aquarids	Jul 29	Jul 19-Aug 8	Jul 15-Aug 29	25	2:00 A.M.	22h30m, 0° and 22h40m, -16°
5. Perseids	Aug 12	Aug 9-14	Jul 23-Aug 20	50	4:00 A.M.	3h4m, +58°
6. Orionids	Oct 21	Oct 20-25	Oct 2-Nov 7	25	4:00 A.M.	6h12.5m, + 13.5° and 6h25m, +19.5°
7. Taurids	Nov 3	Oct 20-30	Sep 15-Dec 15	10	12:00 A.M.	3h32m, + 14° and 4h16m, +22°
8. Leonids	Nov 18	Nov 16-20	Nov 14-20	5	5:00 A.M.	10h8m, +22°
9. Geminids	Dec 14	Dec 12-15	Dec 4-16	50	2:00 A.M.	7h28m, +32°

[a] Period during which shower produces at least one-quarter of its maximum number of meteors per hour.
[b] Approximate number per hour for a well-placed observer with clear, dark skies at time of maximum.
[c] Time (standard or daylight savings) when radiant is highest around date of maximum.
[d] Radiant at date of maximum.

are still years in which they show a dash of their old greatness. For instance, in 1982, an amazing outburst of Lyrids occurred, with rates of 75 an hour and, for a few minutes, even 250 an hour. The radiant is relatively high as seen from midnorthern latitudes by the middle of the night.

Eta Aquarids. This is one of the two meteor showers caused by particles originally released by Halley's Comet at past returns. The radiant doesn't rise until about 2 A.M. and is still rather low in the southeast just before dawn for midnorthern latitude observers.

Delta Aquarids. These may be the most plentiful of all meteors if you add up the total number over the course of several weeks. Although they typically come to a gradual peak in late July, there are usually some left to fire back from the south when the Perseids are shooting out of the northeast and north at maximum around August 12. Actually the later Delta Aquarids belong to a separate branch of the shower; observers may think they are seeing Delta Aquarids when they are actually viewing Alpha Capricornid meteors shooting out of this section of the heavens, mostly around July 30.

Perseids. This is many people's favorite meteor shower because it occurs on comfortable and convenient summer nights and because of its great numbers: only the Geminids often produce more meteors. The Perseids are derived from Comet Swift-Tuttle, which made a surprise return in 1992 after being gone for 130 years. This comet has been called "the single most dangerous object known to humankind" by *Sky & Telescope* magazine, because it is the largest known object that currently has the capability of hitting Earth. It might be much larger than the asteroid or comet that most scientists now believe destroyed the dinosaurs along with three-quarters of all species on Earth sixty-five million years ago. In 1992 there was even a calculation that it had a 1 in 10,000 chance of hitting Earth at its next return, but we now believe we are probably safe from it for at least a few thousand years, and it may never hit at all. You can read much more about this fascinating comet in my book *Comet of the Century* (Springer-Verlag, 1997). But for our purposes here, the interesting thing is that Swift-Tuttle's passage brought with it a new, enhanced peak of the Perseids that seems to happen about half a day before the usual one. As the time since the comet's passage increases, we may see this enhanced peak disappear, but through the mid-1990s it was still producing more meteors for a few hours than at the usual excellent peak. So it is best to go out on several nights (especially before maximum) to observe these exciting meteors. The Perseid radiant is highest around 4 A.M., not long before morning twilight at midnorthern latitudes, but good numbers can be seen earlier as Perseus climbs the northeast sky.

Orionids. This is the other meteor shower derived from Halley's Comet. (The Orionid stream has diffused from the inbound leg of Halley's orbit, May's Eta Aquarids from the outbound leg.) Like the Perseids, they are especially fast meteors, with the radiant highest around 4 A.M. But many of the Orionids are faint, so really dark country skies greatly benefit their observers. Their numbers vary from year to year—sometimes remarkable displays of dozens per hour can be seen. The Orionids also have several peaks, so it is worthwhile watching for them on a number of mornings.

Taurids. This shower extends over many weeks with an uncertain peak date and fairly low rates. Its radiant gets high in the evening, highest in the middle of the night, and it has a reputation for producing fireballs.

Leonids. In most years, on their peak morning, the Leonids produce at best only five to ten meteors per hour for well-placed observers with clear, dark skies. But much more impressive activity is possible for years before and after the passage of their parent comet, Tempel-Tuttle (not to be confused with the Perseids' comet, Swift-Tuttle). In 1996, Leonid rates leaped up to a peak of about thirty per hour, but what was really amazing was how many of these were bright (few, if any, were dimmer than first magnitude) or even fireballs. I watched for four hours before dawn that day and saw three meteors in the magnitude −6 range. These literally lit up the landscape, and two of them left trains visible to the naked eye for well over a minute. Elsewhere, people saw trains lasting much longer, up to forty-five minutes—incredible! But this may be tame compared to what a large part of Earth may get to see on November 17–18 of 1998 and 1999. Comet Tempel-Tuttle passed Earth and Sun, becoming a nice object in telescopes near the North Star in January 1998. And predictions are that in both 1998 and 1999 there's a good chance Earth will pass through the famous Leonid meteor swarm. This is the swarm that we mostly missed around 1900 and 1933 but that produced the greatest recorded meteor storms in history in 1833, 1866, and 1966. In 1966 I was a child and got to see only the spectacular prelude to the big show. Out in Arizona, while most people slept, a group of skywatchers on Kitt Peak watched Leonid rates pick up to tens of thousands per hour and finally, briefly, to a rate estimated as high as 500,000 per hour. One observer said it was like seeing a waterfall of shooting stars pouring down the sides of the sky. If you're reading this book after November 1999, the bad news is that there may be no full-force Leonid meteor storms in the twenty-first century. The reasonably good news is that for many years after 2000 there could still be some displays of Leonids that are far more impressive than usual.

Geminids. Although the Leonids could be sensational for the next few years, the Geminid meteor shower may be the best in most years. Rates of one Geminid per minute may be observed from country locations in good years, and there are typically enough bright Geminids to provide even city dwellers with some thrills. Most of us have to contend with very cold weather to see the Geminids, but at least something close to peak numbers can be seen by 10 or 11 P.M., and the radiant is highest many hours before dawn. The Geminids seem to be the only major meteor shower derived from an asteroid—but astronomers believe that this peculiar asteroid could actually be the extinct core of a comet. Whatever the case may be, it seems that the Geminids are a rather young shower, and some experts think their numbers may decline in the twenty-first century. Better get them now, while they're hot!

Painting of nineteenth century display of Leonid meteor storm over Niagara Falls.

SUMMARY

The best meteor showers in most years are the Quadrantids, Lyrids, Eta Aquarids, Delta Aquarids, Perseids, Orionids, Taurids, Leonids, and Geminids. Each has its own interesting characteristics and opportunities for the observer.

✦ NIGHT 26 ✦

CONJUNCTIONS AND OCCULTATIONS

✦ ✦ ✦

Time: Most nights, but especially the nights of the
best pairings and gatherings of celestial objects.

WHAT COULD POSSIBLY be more beautiful than the Moon, planets, and stars themselves?

These celestial bodies together—close together, in different arrangements that express the intricate interworkings of various celestial movements, arrangements that change by the month, week, night, and hour. Whether you appreciate only the still images—the patterns of Moon, stars, planets at this time or that—or are fascinated also by the celestial choreography that underlies the magic of their dance, there is little doubt that conjunctions and occultations provide us with some of astronomy's most awesome and lovely sights.

Think of the three celestial sights that most thrilled and terrified our superstitious ancestors and seemed most portentous to them: eclipses, comets, and conjunctions.

DEFINING CONJUNCTIONS

The older, looser definition of conjunction is a temporary pairing or gathering of celestial objects in approximately the same part of the sky. But since conjunctions and occultations really are events of exquisitely precise timing and placement (that is part of their thrill and appeal), it's important to use accurate terms in defining and describing them.

Technically, a **conjunction** occurs when one celestial body moves into a position that is at the same right ascension or ecliptic longitude as another—which is to say, depending on your coordinate system, when one object is due north or due south (or, very rarely, right in front of or behind) another. An **appulse** occurs when two celestial bodies come closest together. If two planets are taking almost parallel courses through a constellation, then the moment one planet is due north of the other is roughly the moment they are closest together in the sky. But in many cases, the conjunction and the appulse occur hours or even days apart. In fact, sometimes a planet will pull close to a star or other planet and then, by beginning or ending retrograde motion, reverse direction and pull away before ever passing north or south of the second object. In such a case, there would be an appulse, but no conjunction—at least not in the strict sense of the word.

Incidentally, the Belgian astronomical calculator Jean Meeus came up with two terms for other special arrangements of celestial bodies. When one planet comes to within five degrees of another without ever having a conjunction—a fairly rare event—Meeus calls it a "quasi-conjunction." Likewise, if three celestial objects can all be contained within a circle five degrees in diameter, Meeus calls it a "trio." In each decade there are typically only a few trios of bright planets at a great enough angular elongation from the Sun to see—but such events are well worth the wait!

Other, larger, rarer arrangements of planets are possible, of course. But the Moon moves much faster than the planets, so its conjunctions with bright stars and planets are far more numerous, though short-lived. Every month, the Moon more or less passes by every visible planet and by every bright star in the zodiac. But this may happen when Asia, not America, is in darkness. The Moon crosses the ecliptic at different points each month, and how far it departs from the ecliptic at its extremes for the month varies greatly over the course of many years. (These are all consequences of minor complexities of the Moon's orbit that we haven't discussed.) So the variety of Moon-planet and Moon-star conjunctions is tremendous, with every season and every year bringing different ones.

OBSERVE CONJUNCTIONS

Following the Moon around the heavens during the course of a month is a great way to learn bright planets and stars. Not quite certain which of those points of light is Saturn, or Regulus? Then find out which night the Moon will be nearest, or to the left or right of the planet or star, and the Moon will be your guide. Where can you obtain this information? From sky simulator software. In some almanacs. In the popular astronomy magazines. And, perhaps best of all, in the little diagrams for every night that a notable Moon-star or

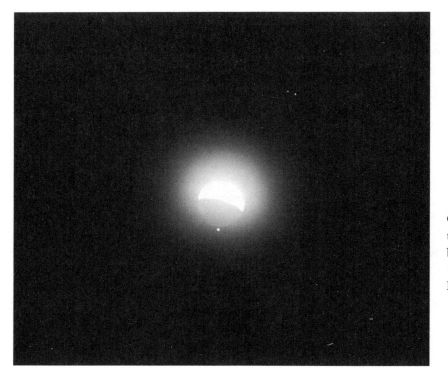

Occultation of the star Aldebaran by the Moon. (Photographed by Kosmas Gazeas)

Moon-planet arrangement occurs—diagrams which appear in the monthly Abrams Planetarium Sky Calendar (see the Sources of Information in the back of this book).

Go out tonight and see if the Moon is in conjunction with any bright point of light, or if any other pairing of celestial objects is occurring. You can tip the balance in your favor by first checking one of the monthly astronomy magazines or the diagrams in this month's Abrams Sky Calendar.

OCCULTATIONS

How close can a conjunction get? So close that one of the celestial objects passes right in front of the other. Such an event, when one celestial object hides another, is called an **occultation.** ("To occult" means "to hide"; "the occult" is supposed to be that which is hidden.)

The only solar system object visible at night that is big enough to cause a lot of occultations is the Moon. But you'd be surprised how seldom the Moon hides a naked-eye star—the naked-eye stars are fewer, the Moon's apparent size smaller, and the sky even bigger than most of us think. Furthermore, when the Moon is anywhere near full, it is so bright that viewing even a first magnitude star right up to its bright edge with the naked eye can be difficult. Lunar occultations are much more frequently visible if you use binoculars or, especially, a telescope. Occultations of stars by planets are exceedingly rare, but when one does occur, a telescope is certainly going to be needed. And though we are not yet to the section on telescopes, our current Night seems like a good place to discuss occultations.

The disappearance of a star behind the moon is called **immersion**. Immersion is much easier to view when the Moon is waxing, because the waxing Moon has a dark eastward limb (edge) against which a star can be seen. When a star or other celestial object comes out from behind the Moon, the event is called **emersion**. You need to know at what position along the Moon's westward limb the star is going to suddenly pop out.

A **grazing occultation** is particularly exciting. In such an event the star is seen to brush along the edge of the Moon, alternately blinking on as it shines through a lunar valley and off as it passes behind a lunar mountain. It's even possible for a star to simply fade, then brighten—indicating that only part of it went behind the lunar limb.

Grazing occultations (or regular occultations at immersion and emersion) can thus provide information about whether a star might be a close double or a very large star— so large that the Moon's motion doesn't hide it instantly. If the object being occulted is a planet, imagine the thrill of using your telescope and seeing part of the crescent of Venus or the rings of Saturn sticking out from behind illuminated mountains on the Moon's edge—or, if it's behind the dark part of the Moon, a piece of Venus or Saturn seeming to hang all by itself in the dark sky. Grazing and regular occultations can help us learn better the topography of the Moon at the precise zone where the star or planet is being occulted. Occultation experts even believe they have obtained evidence of some asteroids having moons of their own by watching a star dim briefly before or after the asteroid occulted. The spacecraft *Galileo* confirmed by direct photography that at least one asteroid, Ida, does have a little moon of its own, Dactyl. As a matter of fact, the rings of Uranus were discovered accidentally when they occulted a star before and after Uranus itself did.

Grazing occultations of a really bright star or planet are rare for any given spot on Earth. For a star, the zone on Earth from which an observer can see a graze is generally only a few miles wide, though it may be thousands of miles long. If you're willing to settle for a fainter star, or to drive a few hours in the night, your possibilities are better. But remember: almost all these events require a telescope. There are exceptions, especially when a bright star or planet goes behind or comes out from the dark edge of the Moon (and preferably a rather thin and therefore not overbearingly bright Moon).

The best places to learn about occultations is in the expert articles on them in the January and February issues of *Sky & Telescope* magazine each year, or directly from IOTA, the International Occultation Timing Association (see Sources of Information).

SUMMARY

Loosely defined, a conjunction is a close meeting of two celestial objects. The stricter definition is that it's when one celestial object passes through the same right ascension (or ecliptic longitude) as another; in other words, it passes due north or due south or centrally in front of or behind the other. There are Moon-planet, Moon-star, planet-planet, and planet-star conjunctions. The moment of closest approach in the sky, in angular distance, between two objects is the appulse, which does not necessarily happen at the same time as the conjunction. A conjunction in which one celestial object at least partly hides another is called an occultation. Most commonly this involves the Moon hiding a star. An occultation in which a star or planet creeps along the Moon's limb (edge), alternately hidden and revealed, is called a grazing occultation.

✦ NIGHT 27 ✦

ECLIPSES AND
SAFE SOLAR OBSERVATION

Time: Whenever an eclipse is visible or, for regular

solar observation, any day the Sun is visible.

NO ONE NEEDS to make an elaborate argument that eclipses are spellbinding. We know they are! Solar and lunar eclipses are the greatest transformations we ever get to see of the sky's two brightest and most prominent bodies. Furthermore, we—that is to say, our own world and all of us on it—are an integral part of eclipses: the Earth must be in line with the Sun and the Moon for an eclipse to occur.

The only major problems for the beginner wishing to learn about eclipses is that they are infrequent and that seeing solar eclipses requires solar observation, which needs to be done properly to be safe.

RARITY OF ECLIPSES

Up until our last few Nights, every kind of celestial object or event discussed was typically available almost any night or at least almost any month if you were willing to go out very late or very early. The Moon, bright stars, planets, constellations, meteors and satellites, the Milky Way, double and variable stars, deep-sky objects: good representatives of each of these classes of heavenly objects can be seen on a regular basis. Not until we learned about meteor showers and certain kinds of conjunctions and occultations did we face the prospect of possibly having to wait as long as several months to make an observation.

It is true that Earth always experiences at least a few eclipses in a year—somewhere on the globe. But there are some years when our particular home location might experience only one eclipse, or even none at all. The year's only observable eclipse might be of the mildest kind, and if the night or day is overcast . . . We are out of luck.

Clearly, we have to accept from the start that eclipses are rather rare (which is part of their allure) and that we will often have to wait quite a long time to see the next one. But we can console ourselves with the knowledge that eclipses are well worth almost any wait.

HOW ECLIPSES OCCUR AND WHY THEY'RE RARE

What are the exact lineups of worlds that produce eclipses—and why don't they happen more often?

Eclipses of the Sun (solar eclipses) happen when the Moon passes between Earth and the Sun. The line in space is Sun–Moon–Earth. Eclipses of the Moon (lunar eclipses) happen when Earth comes between Sun and Moon. The line in space is Sun–Earth–Moon.

If you think back, you'll remember that the Sun–Moon–Earth arrangement is what we have each month at New Moon. And the Sun–Earth–Moon arrangement is what we have

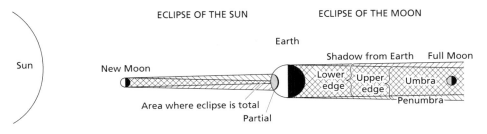

Figure 27.
Solar and lunar eclipse line-up (not to scale and, of course, a solar and lunar eclipse can't occur at the same time).

each month at Full Moon. So why isn't there a solar eclipse at every New Moon and a lunar eclipse at every Full Moon?

We alluded to the reason briefly earlier in the book: the Moon's orbit is tilted. That is to say, the Moon's orbit is tilted with respect to the plane containing Sun and Earth, the plane of Earth's orbit—the plane whose projection on the sky is that midline of the zodiac, the ecliptic. About one-half of each month, the Moon is north of the ecliptic (north of the plane of Earth's orbit); the other half of each month, the Moon is south of the ecliptic. The Moon lies right on the ecliptic at two spots during the month, during each full orbit it makes around the heavens. These spots are the Moon's **nodes**. The node at which the Moon is passing north across the ecliptic is the **ascending node**; the node at which it is passing south across the ecliptic is the **descending node**. Planets too have ascending and descending nodes with respect to the ecliptic, but they go through these nodes far less often than the nearby Moon with its monthly orbit.

A solar or lunar eclipse can only happen when the Moon is at or near one of its nodes. So if the New Moon isn't at or near a node, there will be no solar eclipse; if the Full Moon isn't at or near a node, there will be no lunar eclipse. Another way to explain the situation is this: at New Moon, the Moon usually passes a little north or south of the Sun's disk, and so doesn't eclipse it; at Full Moon, the Moon usually passes a little north or south of the Earth's shadow (which is direct cause of a lunar eclipse), and so doesn't get eclipsed by it.

Another reason why a person at a fixed location doesn't see eclipses every year is that sometimes an eclipse occurs when another part of our planet is facing the Moon. Thus, some total lunar eclipses may be visible in Asia—but not halfway around the world in the Americas, and vice versa. Only half of the Earth is turned toward the Moon at any one time. Both lunar and solar eclipses can last for several hours; during that time the Earth turns, so that more than half the Earth is exposed to the eclipsed Moon or Sun. But, as we'll see later, a *total* solar eclipse is visible over only a very small percentage of Earth's surface, and even partial solar eclipses can be seen from much less than half the Earth.

Solar eclipses are visible from a smaller geographic area than lunar eclipses. Why? To see a lunar eclipse, we only have to be in the hemisphere of Earth that is facing the Moon; to see a solar eclipse, we have to be somewhere within the zone on the Earth that the Moon's shadow passes over. This area is no more than several thousand miles wide, not the eight thousand miles necessary to shadow an entire hemisphere of Earth.

HOW NOT TO OBSERVE SOLAR ECLIPSES

Solar observation is the only activity in astronomy that is, in and of itself, potentially dangerous. I cannot stress strongly enough how important it is to use proper techniques when observing the Sun. NEVER LOOK DIRECTLY AT THE SUN WITH THE NAKED EYE, AND ESPECIALLY NOT WITH ANY BINOCULARS OR TELESCOPE. Even a moment of looking at the Sun directly through a telescope could result in PERMANENT BLINDNESS. No glimpse of the heavens is worth losing your sight.

What about filters? If you're thinking about using some kind of sunglasses or smoked glass or film negatives (people get confused about these things—are you supposed to use undeveloped film or overexposed film or several layers of film negatives?), the answer is ABSOLUTELY NOT. Most of these may greatly reduce the amount of visible light coming from the Sun, but they still let through a lot of infrared and ultraviolet rays. You keep looking at the Sun because it is darkened, and meanwhile the unseen radiation is doing its damage. The retina itself has no pain receptors, so without your even knowing it, a part of your retina burns and you lose—maybe forever—a part of your field of vision.

I, and most astronomy authorities, even recommend against using the special "solar filters" for eyepieces that come with many inexpensive telescopes. These can be poorly made and may crack dangerously. There are Mylar filters that are safe. And for observation with the eye alone—NOT with binoculars or telescope—you can look through welder's glass shade number 14 and be safe. (Don't use any lower number.) I strongly advise the beginner in solar observation to always have careful supervision from an experienced observer.

SAFE SOLAR PROJECTION

Now that I've scared you (and intentionally so!), let me offer a reassurance: there is a simple technique whereby a person can view the Sun safely. It is called **solar projection**, and it's just what its name says: projecting the image of the Sun onto a screen of some kind, like the ground, pavement, a piece of white cardboard, or whatever.

If you have a telescope or binoculars, you must make sure that the image of the Sun is projected away from you (and of course not toward anyone else). How can you get the Sun in a telescope without looking through the telescope? It's actually not too difficult. The trick is to get the tube of the telescope (or tubes of the binoculars) to make the smallest, usually roundest, possible shadow. When it does, the Sun should be shining through your telescope. You should see a much brighter area on the screen in front of your optical instrument's eyepiece. Adjust the focus, or the distance of the screen from the eyepiece, until the edge of the Sun's image is sharp—and until, if you are using a sizable telescope, you are probably seeing sharp images of sunspots! It's thrilling to behold these less hot and therefore seemingly dark regions on the surface of our home star. Imagine now the additional thrill of seeing a huge black bite being taken out of the edge of the solar image and watching the edge of that blackness progress and overwhelm one sunspot group after another—and knowing it is the Moon that is doing this to the Sun in exquisite detail, all on your little homemade cardboard screen.

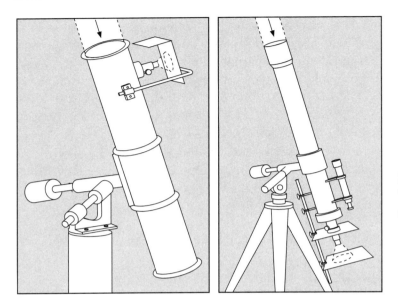

Figure 28.
Telescope setups
for solar projection.

But suppose you have no telescope or binoculars? (We appear to be champing at the bit to use these instruments as we approach the final section of this book—which is devoted to the views that they can provide.) Never fear. You won't easily detect sunspots with the device I'm going to mention, but you will certainly be able to enjoy the progress of a solar eclipse. The device? Two pieces of cardboard with a pinhole in one of them!

Believe it or not, when you hold the pieces of cardboard roughly perpendicular to the incoming rays of sunlight, the bright circle of light that appears on the second piece really is a tiny image of the Sun. Varying the distance between the two pieces of cardboard varies the brightness, size, and sharpness of the image. Experiment to get your best picture.

If you don't have cardboard, you can try to project an image (toward a screen of some kind, even the ground) by creating a chink between your fingers for sunlight to shine through. Or, if there are shade trees in leaf nearby, look at the dappling of sunlight beneath them: you should see multiple big images of the Sun (though they are likely to be elliptical rather than round), each with a bite out of it—an image of the solar eclipse.

SUMMARY

A lunar eclipse occurs when Earth gets between Sun and Moon and casts its shadow over the Moon. A solar eclipse occurs when the Moon gets between Sun and Earth and casts its shadow on Earth. Eclipses are fairly rare because most months the lineups of Sun–Earth–Moon at Full Moon and Sun–Moon–Earth at New Moon are not precise enough to cause a lunar or solar eclipse, respectively. For an eclipse to occur, the Moon must be quite near to its ascending or descending node, the places where it passes north or south across the ecliptic, the plane of Earth's orbit. The Sun, during a partial eclipse or when there is no eclipse in progress, can be observed safely with the naked eye by looking through shade number 14 welder's glass or a well-designed Mylar solar filter. DO NOT USE OTHER METHODS OR FILTERS TO OBSERVE THE SUN WITHOUT EXPERT GUIDANCE! With binoculars or telescope the easiest safe solar observation is done by solar projection of the Sun's image onto a piece of cardboard or other screen, but extreme caution MUST be taken to do solar projection correctly. Solar projection can be done with "equipment" as simple as two pieces of cardboard, one with a pinhole in it, or even with the leaves of a tree, with patches of sunlight dappling the shaded ground below.

◆ NIGHT 28 ◆

ECLIPSES OF THE SUN

◆　◆　◆

Time: Any day that there is a solar eclipse (see Appendix 3
for listing of upcoming solar eclipses).

IN THE PRECEDING pages, I discussed how solar and lunar eclipses occur and why they are relatively rare. I also discussed the dangers of trying to observe the Sun—during an eclipse or otherwise—and described the few safe methods for solar observation. Now it is time for us to look at the different types of solar eclipse and to learn how they can affect the whole sensory environment of land, sea, and air around you—especially if the event is that masterpiece of the heavens, the shatteringly beautiful phenomenon we call a total eclipse of the Sun.

THE THREE MAJOR TYPES OF SOLAR ECLIPSE

The amount of darkening you experience at a solar eclipse varies tremendously depending on the type of eclipse. Some partial solar eclipses are so slight that even an expert observer can't detect a change in the brightness of the sky or landscape. On the other hand, a total eclipse of the Sun gives me a rationale for devoting much of this Night to solar eclipses, which technically can only be observed during the day, when the Sun is above the horizon, not at night. The darkness caused by a total eclipse of the Sun really is like the falling of night, even if it occurs at noon!

What are the three major types of solar eclipse? A **partial solar eclipse** occurs when only part of the Sun's blinding disk is covered by the Moon. A **total solar eclipse** takes place when the Sun is completely covered by the Moon. Simple enough, right? But there is a third major variety. You could categorize it as a subclass of partial eclipses, but it is so special that it deserves its own class. An **annular solar eclipse** occurs when the Moon is so far from Earth that its form does not quite cover the entire solar disk, even if the Moon passes right in front of the Sun. Annular means "ringlike" (from Latin *annulus,* "a ring") for indeed that's what is left uneclipsed in the middle of an annular eclipse: a narrow ring of the Sun's blazing surface shines around the silhouetted night side of the Moon.

Much has been made about the strange coincidence that the Sun and the Moon appear almost identical in size as seen from Earth. But the Moon's distance from Earth varies so it appears about ⅛th less wide when farthest than when closest. Only when the Moon is at least a little closer than average is its apparent size great enough to block out all the dazzling part of the Sun and produce a total, not just an annular, eclipse.

EXPERIENCING A SOLAR ECLIPSE

It's not just the Sun and Moon that are interesting to witness during a solar eclipse. You can look, listen, and feel many changes in the sky and landscape, and do so without optical aid.

Set your screen up for solar projection and notice the first bite out of the solar image. The Moon has begun moving in front of the Sun. As the Moon slides one-third or perhaps one-half of the way across the Sun, keep monitoring the brightness and color of the sky and landscape. If you are not convinced at these early stages (which are all that ever occur at small partial eclipses) that you are seeing a darker, bluer sky and a dimmer landscape with a kind of yellowish cast to it, you can check your camera's light meter. And if the eclipse is going to get deeper, get ready for more wonders.

If the Moon is now about two-thirds of the way across the Sun, the dimming of the sky and landscape and the yellowish or yellow-reddish hue to everything is usually unmistakable. Whereas before the light strongly glittered off windows or the chrome of cars, it now glitters with more localized silvery gleams. If you have a thermometer, you will probably see that there has already been a drop in temperature, even if it is morning and the temperature should be going up. (Exactly what the temperature does depends on the time of day and current weather pattern, but drops of as much as 20°F are not rare by the time of total eclipse.)

TOTALITY

Unless you're lucky and this is the once in over three hundred years (on an average) that a total solar eclipse is occurring in your home town, or unless you've traveled (usually thousands of miles) especially to a location for "totality," you have to settle for a large partial (or annular) eclipse. A total eclipse occurs in a band thousands of miles long but only many dozens of miles wide because that is the zone of totality traced out by the moon's central shadow or "umbra."

I will end this Night by describing a total eclipse of the Sun and by urging you do whatever it takes to see at least one in your lifetime.

At some total solar eclipses, you can actually see the shadow of the Moon approaching at great speed as a purple, stormlike wedge of darkness, and long narrow "shadow bands" moving across the landscape. The latter are caused by the twinkling of the Sun when only a tiny thin sliver of it is left. As the front edge of the Moon's shadow reaches the Sun, a darkness like that of deep twilight or a Full Moon night descends, but most of it happens so fast, in the few seconds it takes to cover the last sliver of Sun, that it feels more like being plunged into instant midnight. Now you can look safely, without any eye protection. Bright planets and stars appear here and there in the sky; the sky may look as if it's ringed in orange or red twilight all the way around the horizon, and from around the jet-black disk of the Moon has leaped the "crown" of the eclipse: the pearly white "solar corona," the outer atmosphere of the Sun, has blossomed like flower petals of light.

The inner corona of the Sun is visible in this short photographic exposure of a total solar eclipse taken from the Skylab space station in 1973.

You will have no more than a few minutes of totality. You might be too rushed to see the orange or red tufts of fountainlike "prominences" sticking out from behind the Moon. You may miss at totality's start or end the sight of "Baily's beads"—large dots of uneclipsed solar surface shining through lowlands on the Moon's edge—and the spine-tingling "diamond ring effect," one small speck of uneclipsed Sun shining like a blazing star on the still visible band of the inner corona. But you will have been able to see what you did without any protection to your eyes—even telescopes can be used—during the total part of the total solar eclipse. And you will feel, like other people before you, that a total eclipse of the Sun is the most staggeringly awesome experience in nature that science can predict.

SUMMARY

There are three most important kinds of solar eclipse. A partial solar eclipse occurs when the Sun's blinding disk is only partly hidden by the Moon. A total solar eclipse happens when the Moon hides all of the brilliant Sun from our view. An annular solar eclipse takes place when the Moon passes more or less centrally in front of the Sun but is far enough from Earth that it appears a tiny bit smaller than the Sun. Instead of being totally eclipsed, a thin ring ("annulus") of the Sun's surface remains visible—and remains dangerous to look at directly.

When a solar eclipse hides half of the Sun, observers begin to notice not only a darkening of the sky but also other changes. The landscape changes color because of the sunlight coming from the more reddened edge of the Sun, the temperature drops, and there are changes in cloud cover, wind, and the behavior of animals; for instance, birds, if not utterly confused, go to roost. Just before and after the total stage of a total eclipse, you may see the Moon's shadow approaching or receding and "shadow bands" racing across the landscape. A total eclipse of the Sun is the most staggeringly awesome of all predictable sky events. Light levels drop precipitously, planets and stars may come out in the sky, and many awesome sights, such as Baily's Beads, the diamond-ring effect, and the band of reddened light around the horizon, may be witnessed.

◆ **NIGHT 29** ◆

ECLIPSES OF THE MOON

◆ ◆ ◆

Time: During a lunar eclipse (see Appendix 4
for listing of upcoming lunar eclipses).

A LUNAR ECLIPSE is beginning! At first you notice only a slight shading of the Full Moon on its left limb. But eventually a tiny inky spot appears on that side of the lunar orb. Over the course of an hour or so, the curved edge of that shadow advances with eerie slowness across more and more of the Moon. (With the help of binoculars or a telescope, you can watch the shadow's edge creep up on one bright crater or mountain after another and bury it in darkness.) The curved edge of the Earth's shadow on the Moon is proof, as the ancient Greeks recognized, that the Earth, whose shadow this is on the Moon, is round. By the time this shadow is most of the way across the Moon, you may begin noticing a hint of red in the shadowed part of the Moon. And when the last bright rim is hidden, you will often see color—usually some shade of red—covering the entire greatly darkened Moon. If you are well away from city lights, you will see that the eclipse has brought out multitudes of stars too dim to be seen plainly before the eclipse darkened the blazing Full Moon. Finally, after maybe thirty, sixty, ninety—or even more— minutes of total eclipse, a bright slender slice of the Moon's right (west) limb materializes, perhaps golden trimming for the still visibly reddish rest of the Moon. But before long the amount of yellow Moon outside the Earth's central shadow begins to grow large, and the lunar eclipse goes through its previous stages in opposite order, eventually bringing the eclipse to an end.

THE THREE MAJOR TYPES OF LUNAR ECLIPSE

There are three major types of lunar eclipse, and for the observer there is a big difference among them. To understand the difference, we have to be students of shadow—the Earth's shadow in particular, but also shadows in general.

Using a localized light source like a lamp, hold your hand—or a pencil, or a ball, or virtually any opaque object—in front of a surface on which you can see the object's shadow. You'll find that the shadow consists of a dark inner part and at least a narrow edge that is less dark. The dark inner part is called the **umbra** (from the Latin for "shadow" or "shade"—an umbrella is a "little shade"). From within the umbra an observer's view of the

Eclipse of the Moon, showing the curved edge of Earth's shadow. (Photo by Steve Albers)

light source is completely hidden. The lighter, outer part of the shadow is called the **penumbra** (from Latin for "almost" and "shadow"). From within the penumbra an observer's view of the light source is partly hidden.

The Earth's umbra is a tapering cone that extends over a million miles into space. It is bordered all around by the Earth's penumbra. But the Moon is never more than about a quarter of a million miles from Earth. How wide is the umbra and how wide the penumbra at this distance? The umbra is about twice as wide as the Moon at the Moon's distance from the Earth, and the bordering penumbra is about one more Moon-width all around the umbra. If we imagine a cross-section of the Earth's shadow out at the Moon's distance, it looks as it does in Figure 29.

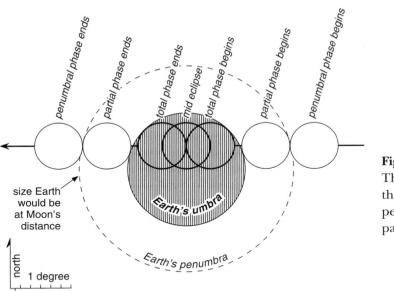

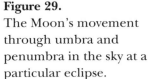

Figure 29.
The Moon's movement
through umbra and
penumbra in the sky at a
particular eclipse.

Which type of lunar eclipse we see depends on how far into the Earth's shadow the Moon penetrates. If the Moon passes only through the penumbra, the event is a **penumbral lunar eclipse**. The penumbra is of such a light shade that the Moon generally must be at least halfway into the penumbra before its most advanced edge begins showing even a slight "stain." A **partial lunar eclipse** occurs when the Moon enters the umbra but only part of the Moon gets covered with it. The first touch of the umbra on the Moon is normally visible as a quite dark notch on the Moon's edge. Finally, if the entire Moon enters the umbra, a **total lunar eclipse** occurs. Every total lunar eclipse has a penumbral and a partial eclipse built into it. And during a total lunar eclipse, we not only see the entire Moon dramatically darkened, we are also treated to a lovely, surprising sight: a reddening of the Moon, with touches of other color.

THE DARKNESS AND COLOR OF TOTAL LUNAR ECLIPSES

Why does the Moon redden during a total eclipse of the Moon? The answer is truly surprising.

The reason why there is any color within the Earth's shadow is that Earth has an atmosphere. The solid body of the Earth blocks direct sunlight from reaching the Moon. But our atmosphere actually refracts—that is, bends—sunlight into the shadow from around the rocky bulk of the planet. In a sense, what we are seeing is the light from every sunrise and sunset happening in the entire world at the time of the lunar eclipse. The Sun looks reddish and dimmed when low in the sky because its light has to traverse a longer pathway through the absorbing and scattering atmosphere than when it is high. But the sunlight that reaches the Moon during a lunar eclipse has to travel even farther, for it exits back out through the atmosphere. It is only reddened slightly more by its final trip—back into our atmosphere after being reflected off the otherwise dark Moon—as long as the eclipsed Moon appears reasonably high in our sky.

Another way to understand this colorful drama is to imagine what a total eclipse would look like to an astronaut on the Moon. First, our astronaut would see the silhouette of the huge Earth start to pass in front of the Sun. Since the Earth would appear about four times wider than the Sun, you would think that sunlight would be completely blocked once the Earth's edge had covered the Sun; at the moment this happened, a wave of darkness—the forward edge of the Earth's umbra—would rush across the lunar landscape around our astronaut. But darkness would not be complete. Instead of the Earth appearing as merely a black absence of stars in the sky, a band of luminous red would begin growing. By totality, it would form a complete ring around the Earth's huge form.

THE DANJON SCALE AND DARK ECLIPSES

What makes a total lunar eclipse even more fascinating is that we can never predict exactly how dark it will be or what shade of red and traces of other color will appear on the Moon. Why the variety? It's partly a result of how deeply within Earth's umbra the Moon passes: the nearer to the center of the shadow the Moon goes, generally the darker the eclipse is. But also very important is the condition of the atmosphere along Earth's sunrise-sunset line. If there is heavy cloud in the key regions, not much sunlight will get through. But this by itself will not necessarily make for a dark umbra and dark eclipse. The only times when the Moon turns a really deep blood red or dull gray or even blacks out are

TABLE 4

Danjon's Brightness Scale for Total Lunar Eclipses

L = 0	Very dark eclipse; Moon hardly visible, especially near midtotality
L = 1	Dark eclipse; gray-to-brown coloring; details on the disk hardly discernible
L = 2	Dark red or rust-colored eclipse with dark area in the center of shadow, the edge brighter
L = 3	Brick red eclipse, the shadow often bordered with a brighter yellow edge
L = 4	Orange or copper-colored, very bright eclipse with bright bluish edge

L stands for luminosity.

when the upper atmosphere along the sunrise-sunset line is also blocked. And the only thing that can clog the atmosphere that high (ten, twenty, thirty miles) is the silicate particle cloud or, more effectively, the sulfuric acid haze cloud that can be produced by certain kinds of great volcanic eruptions.

A powerful sideways eruption at a fairly high latitude—such as that of Mount Saint Helens in 1980—will not necessarily put huge enough quantities of sulfur dioxide gas up into the stratosphere. But eruptions like those of El Chichon (Mexico, 1982) and Pinatubo (the Phillipines, 1991) have placed millions of tons of sulfur dioxide high enough to get swept around the world and to start combining with water vapor to become an extremely high-altitude sulfuric acid haze. That haze got illuminated by sunlight long after sunset (and before sunrise) came at ground level, producing long-lasting and flamingly bright and colorful twilights. But very little light got through to exit back out of the atmosphere into Earth's umbra and reach the eclipsed Moon. Total lunar eclipses in the year or two after the eruptions were rare dark ones in which the Moon virtually disappeared around mideclipse. Another such eclipse was that of December 30, 1963, after the eruption of Mt. Agung.

More typically, the Moon when totally eclipsed is a brighter orange or pink, still shining as brightly as Venus or Jupiter, though the light is spread over a much larger area than that of the naked-eye planets. You can contribute a bit to science, as well as enjoy yourself, by trying to rate the brightness of the totally eclipsed Moon on the **Danjon Scale** (see Table 4). You can report your rating and other aspects of your observation to your favorite astronomy magazine (see Sources of Information). Notice that each brightness level on the Danjon Scale comes with its characteristic colors. But this is only a general guide. You will eventually see just about every color imaginable during some lunar eclipse. Of course, some of the more subtle hues are better detected by telescopic observation. But plenty of colors can be glimpsed with the naked eye. And the naked-eye view offers those incredible vistas of the strange Moon in its environment of starry sky and landscape that optical instruments with narrower fields of view can't supply.

SUMMARY

There are three important kinds of lunar eclipse. In a penumbral eclipse, the Moon only passes through the penumbra, the less dark outer shadow of Earth. In a partial eclipse, the Moon enters partway into the umbra, the dark central shadow of Earth. In a total eclipse, the Moon enters entirely into the umbra. The Moon usually takes on a reddish hue during total eclipse; the color is caused by sunlight that has been refracted (bent) into Earth's umbra by Earth's atmosphere. The color is deeply reddened by the long grazing passages through the atmosphere and the less grazing passages back into the atmosphere after being reflected off the Moon. Depending on how much the atmosphere is blocked by cloud or volcanic haze around the key sunrise-sunset line of Earth, and also depending on how deeply into the umbra the Moon goes, the Moon can appear any shade, from a very bright orange to a deep red, gray, or even black (disappeared) during the total eclipse. Observers can rate the darkness of the eclipse on the Danjon Scale. Among the other beauties of lunar eclipses is the appearance of numerous stars as more and more of the Full Moon is eclipsed.

PART III

NIGHTS OF THE HEAVENS IN DEPTH

✦ NIGHT 30 ✦

THE MOON'S CHANGING APPEARANCE AND "SEEING"

✦ ✦ ✦

Time: Whenever the Moon is in the sky. Certain phases
are better than others for noticing certain aspects
of the Moon's appearance.

WE NOW COME at last to the telescopic section of this book. If you don't have a telescope, many of the sights can be seen—sometimes surprisingly well—with binoculars. You may also be able find a nearby astronomy club, college, or museum that has a telescope accessible to the public. Or, if all else fails, read on, taking notes about what to do when you do get a telescope.

If you do have a telescope but are just learning to use it, Appendix 12 provides some fundamentals about its use and care.

What is the first telescopic target we will explore here? There's no doubt about that.

THE MAGNIFICENT MOON

The Moon is by far the easiest and most tempting target for a beginner to turn a telescope on. And it is the only world other than our own that, when viewed through binoculars, reveals any structure: in fact, the Moon has hundreds of features. For most people, the view of the Moon through a telescope remains the most breathtaking of all. It's not just the abundance of detail that thrills. It is the starkness of light and dark, the white and yellow and gray and black. And it is the almost tangible textures: the topography, sometimes magnificently rugged; the lit horn of a peak isolated by darkness; multiple wrinkles in the gray smooth rock of a lunar "sea"; impossibly sharp and complex rows of shadows; a clump of small craters like bubbling foam frozen in rock—and the knowledge that on no other night will it look exactly the same.

Before we examine the features on the Moon's surface with optical aid, it's important to understand some general facts about the observability of those features and how they change. The rest of this Night is devoted to this endeavor.

THE TERMINATOR

Part of why the Moon looks different every (Earth) night—even every hour—is that we see on its face the dramatic effects of changing light. Unlike Earth, where the atmosphere scatters light, on the airless Moon everything not in light is in jet blackness and vice versa. Shadows seem to take on a life of their own, and they outline features with exquisite precision. But what you see depends on how high the Sun is in the sky as seen from that particular location on the Moon.

The best time of the lunar day to observe most topographic features on the Moon is near sunrise and sunset—in other words, when the Sun appears low in the sky as seen from that place on the Moon, and even the terrain's least variation in height is outlined by a shadow large enough to see. The line that separates night from day on the Moon or other astronomical body is called the **terminator**.

On the waxing Moon, between New Moon and Full Moon, the part of the terminator running down the side of the Moon facing us is the sunrise line; on the waning Moon, between Full Moon and New Moon, it is the sunset line.

*Go out after nightfall around first quarter and point your telescope or binoculars at the Moon. The views around the **limb** (edge) of the Moon are more washed out with bright light than those near the terminator. You also notice something else about craters and other features near the limb: they appear foreshortened. In other words, the craters, which are more or less round near the middle of the Moon, appear increasingly elliptical and elongated the closer to the limb you look. You are obviously seeing an effect of perspective, the craters near the limb looking as they do because the surface of the spherical moon is curving away from us steeply there.*

About a week later, go out again when the Moon is full. In any but a quite small telescope you will find its brightness overwhelming; it actually begins to cause afterimages and hurt the eyes. And now almost all the Moon's surface seems washed out. The reason is that the Sun is high in the sky from almost everywhere on the earthward face of the Moon, and few shadows are visible. The only exception is near the limb, where just before Full Moon, the Sun is only starting to rise. But the limb region of the Moon curves away from us so much that the features there cannot be seen well at any time, no matter what the lighting.

We noted earlier that Full Moon is not a good time for observing most of the heavenly objects, because they are dim and require a dark sky to be seen well. Now it turns out that Full Moon is not even a good time for observing the Moon! (Actually, though, there are a few lunar features that do look best at Full Moon. Rays, which we'll meet in our next Night, are an example of this.)

THE SAME FACE

It soon becomes clear, even to the careful naked-eye observer, that the Moon always shows approximately the same face to the Earth. To many people, this suggests that the Moon, though we know it goes around the Earth in orbit, does not turn around on its axis at all— if the Moon spun, wouldn't we sometimes see its other side aimed toward us? But this apparent contradiction is easily explained. We have only to remember that if the Moon didn't spin at all, we would see opposite sides of it when it was on opposite sides of its orbit from us.

No, the Moon does spin. It's just that it makes one turn in exactly the amount of time it takes to make one orbit around the Earth. This is called **synchronous rotation** and seems to be quite the rule in our solar system. Almost all the moons of other planets studied by space probe have been shown to exhibit this type of rotation. It appears to be the stable situation a moon arrives at if nothing tremendous disturbs it.

But it is not easy to visualize this setup. The best way to learn it is probably by staging a demonstration with three people. One person plays the role of the Earth. The second person, playing the Moon, faces the first and begins to walk around him or her, making sure to always keep facing the first person. The third individual observes from some distance away. This person sees the "Moon" always facing the "Earth," but can also see the back of the "Moon's" head about half the time, proving that the "Moon" *is* rotating. Try it!

ALMOST THE SAME FACE

We were just discussing how the Moon's appearance keeps changing—but if the Moon always keeps the same hemisphere pointed at us, why doesn't it look exactly the same at the same phase each month?

What I wrote a few paragraphs back was that the Moon always shows *approximately* the same face to Earth. In reality, the face of the Moon pointed toward us nods—a little up and down, a little from side to side. This motion is known as **libration**.

The up-and-down libration occurs because, as we've already learned, the Moon's orbit is tilted—sometimes the Moon is farther north or south than at other times. The side-to-side libration occurs because the Moon's rotation rate is almost steady but the Moon's orbital velocity varies considerably. We've already noted that the Moon at its closest to Earth—a position called **perigee**—is about one-eighth closer to Earth than the Moon at its farthest— a position called **apogee**. Just as a planet closer to the Sun has to move faster to stay in orbit (outward-pulling centrifugal force balanced against the inward-pulling gravity of the Sun), so must the Moon move faster when it is closer to Earth. The orbital motion gets ahead of the rotational motion, and we get to peek around one side of the Moon a bit. So pronounced is this libration that you can detect it even with the naked eye over a period of days. *Look at a waxing crescent Moon with the naked eye and notice that the lunar feature called the Mare Crisium is quite a distance from the lunar limb. Then, many nights later, look again. Now the Mare Crisium is very close to the limb. Whereas before it was elliptical in shape, now it is so foreshortened by libration nodding it toward the limb that it appears as little more than a thick line. The true shape of the Mare Crisium, determined by photos from spacecraft passing directly over it, is quite circular.*

There are a few other minor forms of libration that contribute to the Moon's complicated nodding. With such forces at work, it's little wonder that the Moon and its features never look exactly the same.

"Seeing"

There is a final very important reason why lunar features look different from night to night, even from one moment to the next. You will notice that the sharpness of the lunar features you see varies, sometimes in a matter of minutes or seconds; generally they are much sharper when the Moon is high than when it is low in the sky. The cause for these striking changes has nothing to do with conditions on the lunar surface, or the angle at which sunlight is hitting a feature. In fact, the same changes occur in details of a planet's cloud patterns or the crispness of a star's image.

What causes these effects is variation in the steadiness of Earth's atmosphere, through which you are observing the Moon or other celestial object. This sharpness or fuzziness of astronomical images as a function of how much turbulence of atmosphere you are looking through is called "**seeing**."

Bad "seeing" (shaky, blurred images) can occur even when there is no wind at the surface. It can occur miles above ground level or nearby (even right in your telescope tube) and can be caused by the jet stream far above or our warm bodies up close.

"Seeing" tends to be bad low in the sky where celestial light must travel through a longer pathway of air to get to us, therefore encountering a greater total amount of turbulence. Good "seeing" is generally found near the centers of high pressure systems, bad "seeing" where the jet stream is or where isobars pull close together on a weather map. Be sure not to mistake a badly aligned optical system for bad "seeing."

Summary

The terminator is the line separating day from night on the Moon or on any planet or satellite; the limb is the edge of the Moon or other celestial object. Shadows are sharpest and the greatest amount of detail is visible near the terminator; far from the terminator, the landscape is washed out with sunlight. Almost all of the earthward face of the Moon is washed out by sunlight when the Moon is full. Close to the middle of the Moon, craters and other lunar features appear the proper shape (craters are mostly circular, of course), but the closer to the limb they are, the more foreshortened they get from our perspective (viewing them from the side instead of straight on). The Moon always keeps approximately the same face toward Earth because of its synchronous rotation: its rotation period is the same as its orbital period. Because of processes like libration, the Moon doesn't keep exactly the same face toward Earth. Libration is a nodding of the Moon. Its side-to-side component is caused by variations in the orbital velocity of the Moon around perigee and apogee, which permit us to see a little farther around the leading (left) edge and trailing (right) edge of the steadily rotating Moon.

✦ NIGHT 31 ✦

THE MOON'S MANY FEATURES

✦   ✦   ✦

Time: Any time the Moon is up. Certain times are preferred
for observing certain kinds of lunar feature.

THE MOON'S MANY features include maria (the "seas"), craters, mountain ranges, rays, rills, and other valleys. Most of these are best seen when near the terminator—except for **rays**, the streaks of bright lunar soil radiating out from some craters. *Using Figure 30, find as many of the numerous features listed below as possible.*

THE LUNAR SEAS

They are prominent enough to be seen by the naked eye, appearing as numerous dark patches on the face of the Moon. They are the lunar "seas" or, in Latin, the **maria** (singular **mare**). Spacecraft exploration of the far side of the Moon has shown that most of the

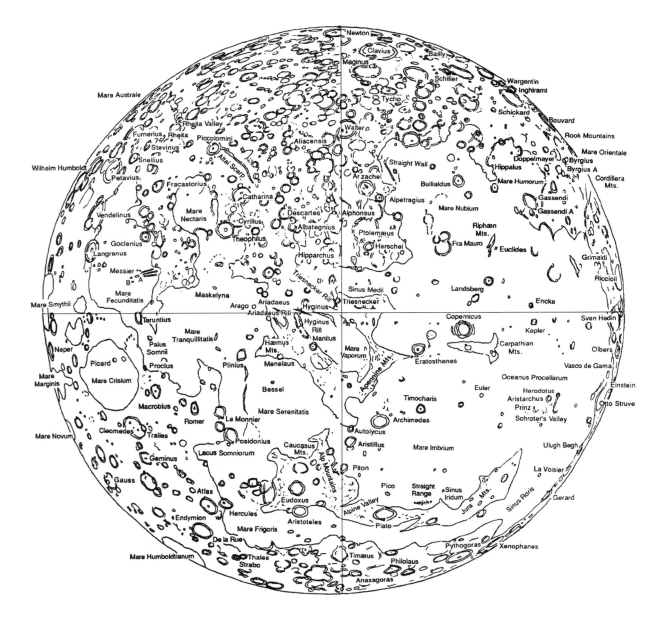

Figure 30.
Map of the Moon.

maria are concentrated on the near side, facing Earth, though no one is yet sure why. We do know, however, what the maria really are: dark plains of hardened lava, produced when asteroid-sized objects collided with the Moon in the early days of the solar system, making wounds up to hundreds of miles wide into which molten rock poured.

Some of the minor maria have names beginning with sinus ("bay"), lacus ("lake"), or palus ("marsh"). The largest of all maria is an oceanus ("ocean"). But these are all essentially similar landforms, contrasting with the brighter highland regions of the Moon. Here is a brief selection of the most interesting maria, in roughly the order you will encounter them in the course of the lunar month.

The *Mare Crisium* (Sea of Crises) is the only important mare on the Moon's near side that is isolated from the others, surrounded by bright highland. As discussed in the previous Night, the Mare Crisium is a handy guide to libration on the waning crescent and up until Full Moon. Several rays from the bright crater Proclus cross Mare Crisium when the Sun gets high over it. *Mare Fecunditatis* (Sea of Fertility) contains the large crater Goclenius and the paired craters Messier and Messier A (Messier A has twin rays going out from it). *Mare Nectaris* (Sea of Nectar) is small and has the bay known as Fracastorius at one end of it. *Mare Tranquillitatis* (Sea of Tranquillity) is large and well formed, and connects with four other maria. But it will perhaps forever be best known as the landing spot of Apollo 11, the mission that on July 20, 1969, put the first men on the Moon. *Mare Serenitatis* (Sea of Serenity) is large and almost round, with no large craters on it—but as the Moon waxes toward full, a bright ray crosses it that seems to be traceable hundreds of miles back to the great crater Tycho.

After first quarter, sunlight hits the most majestic of all the lunar seas, *Mare Imbrium* (Sea of Showers). It is bounded by five mountain ranges and has on it three of the Moon's most striking isolated peaks and ridges. Great craters spot Mare Imbrium, and great craters neighbor it. Attached to Mare Imbrium is the lovely *Sinus Iridum* (Bay of Rainbows), with its one cape looking like a jeweled handle at sunrise and its other cape in the form of "the Moon Maiden" jutting out into Imbrium.

Imbrium and the following maria can be observed as the time of Full Moon nears, but can also be enjoyed on the waning Moon, and experienced at sunset when the Moon is a waning crescent. *Mare Nubium* (Sea of Clouds) and *Mare Vaporum* (Sea of Vapors) are less interesting, though the latter does have rills (see below) running in and to it. *Mare Frigoris* (Sea of Cold) is the skinny (partly because it's foreshortened) mare that extends most of the way across the high northern region of the Moon's earthward face. *Oceanus Procellarum* (Ocean of Storms) is the largest of all the maria, but its borders are ill defined and its patchy floor is upstaged by the mighty craters on or near it. *Mare Humorum* (Sea of Moisture) is small but has two big bays and is bordered to the north by the great crater Gassendi.

CRATERS

Many lunar craters are large enough or otherwise prominent enough to be glimpsed in binoculars, but the magnification of a telescope is really required for detailed views. Let's consider just a few of the most spectacular of craters and their interesting neighbors.

- *Archimedes, Aristillus, and Autolycus.* These three magnificent craters are found together in Mare Imbrium. Aristillus has a three-peaked central mountain mass and walls as high as 11,000 feet. In contrast, Archimedes has much lower walls, a dark floor, and no central mountain.
- *Aristarchus and Herodotus.* Aristarchus is the brightest of all craters, though it's not extremely large. Its neighbor, Herodotus, is remarkably darker; near its north wall is the unique Schroter's Valley.
- *Clavius and Bailly.* These largest craters (of the variety called "walled plains") on the Moon's near side are poorly placed in the far south of the Moon, but the slightly smaller Clavius is farther from the limb, thus easier to see well.
- *Copernicus and Erastosthenes.* Anywhere else on the Moon the large, deep, beautifully formed Erastosthenes would be the showpiece of the region. But Copernicus is the Moon's most magnificent crater: it's 17,000 feet deep from wall rim to floor in places, and its incredibly intricate interior structure includes three huge central mountain masses and huge areas of landslide debris. Copernicus is tremendously bright with a mighty ray system and a splendid maria background to isolate it.
- *Kepler.* Though much smaller than Copernicus, Kepler is very bright, very detailed, and has a great ray system that intermingles with that of Copernicus around Full Moon.
- *Grimaldi and Riccioli.* The Moon's darkest craters (especially Grimaldi) are both located at the west (the side that faces east on Earth's sky) limb of the Moon, so they only come into view just before Full Moon. They are best observed on the waning Moon. Riccioli is very near the limb.
- *Hipparchus, Ptolemaeus, and Alphonsus.* This line of giant, worn craters is near the Moon's central north-south meridian. Observe when the terminator is near; otherwise they are hard to see.
- *Langrenus, Petavius, and Vendelinus.* Petavius is the greatest of this roughly north-south line of mighty craters on the thin waxing crescent.
- *Plato.* One of the Moon's most prominent craters due to its combination of size, darkness, and setting, Plato is on a fairly thin strip of highland between Mare Imbrium and Mare Frigoris, not far from the most glorious regions of Imbrium.
- *Proclus.* The rays of this small but very bright crater pass across Mare Crisium when the Sun is high.

- *Theophilus, Cyrillus, and Catharina.* Perhaps only Copernicus is more magnificent than Theophilus with its extraordinary detail and depth (as much as 18,000 feet between rim and floor). The other craters in its chain, Cyrillus and Catharina, are far less impressive, but just beyond the latter begins the long and prominent Altai Scarp. All these features are best seen when the Moon is about six days old.
- *Tycho.* Although not in itself the grandest of the Moon's craters, Tycho, with its incomparable ray system, is the most prominent feature around Full Moon. Some of the rays reach for hundreds of miles.

MOUNTAINS, RILLS, AND MORE

The Moon's greatest mountain range is the *Apennines,* which extends for 600 miles with peaks up to 18,000 feet high. Just after first quarter, one end of the Apennines catches sunrise before the lowlands around it and typically forms a deformation in the terminator noticeable even with the naked eye. Mare Imbrium is bordered by five mountain ranges, among them the *Jura Mountains,* which stick out as a cape to form at sunrise the jeweled handle of Sinus Iridum. The lunar *Alps* are split for eighty miles by the dramatic *Alpine Valley.*

Pico and *Piton* are isolated peaks on Mare Imbrium that stand out as spots of light at lunar sunrise and cast magnificent long shadows soon after. Not far from them is a perfectly formed forty-mile-long ridge almost 6,000 feet tall, the *Straight Range.* The Straight Range is not to be confused with a stranger and even more thrilling formation near Mare Nubium: the *Straight Wall.* The Straight Wall looks like a cliff but is technically a fault. It is seventy-five miles long, over a thousand feet high, and fairly steep. When the Moon is about nine days old, the Straight Wall looks like a dark line. It disappears altogether for several days around Full Moon. But when it reappears around last quarter, its westward-facing surface is brightly lit by the Sun and it looks like nothing else on all the Moon: a brilliant scratch.

In addition to the *Altai Scarp,* mentioned above, and the *Rheita Valley,* which is really a long chain of craters, the Moon offers long, usually narrow and meandering valleys called **rills**. The most amazing is Schroter's Valley (also mentioned above), but other famous rill systems are those of Hyginus, Ariadaeus, and Triesnecker.

SUMMARY

The large, dark markings on the Moon are the lunar "seas" or maria, vast plains of volcanic rock. Mare Imbrium is the most visually impressive of the maria. Great craters of the Moon include Copernicus, Kepler, Plato, and Theophilus. Tycho has by far the moon's most magnificent system of rays. Rays are streaks of bright lunar soil thrown out from relatively young

Close-up of
Hadley Rill.
(NASA
photograph)

lunar craters. Other spectacular lunar features include mountain ranges like the Apennines, isolated mountains like Pico and Piton, the Straight Range, valleys like the Alpine Valley, and odder features like the Straight Wall, the Altai Scarp, and the Rheita Valley. Rills are
narrow and often meandering ravines on the Moon. A famous example is Schroter's Valley.

◆ NIGHT 32 ◆

THE GLOBES, MOONS, AND RINGS OF JUPITER AND SATURN

◆ ◆ ◆

Time: Any time that Jupiter or Saturn is visible.

THEY ARE THE MOST dependable planets for observers: the two giants, Jupiter and Saturn. Visible for about eleven months of every year, both Jupiter and Saturn change their
position in the zodiac slowly and always present at least a few features to see in medium-size

or even some small telescopes. The outstanding features of Jupiter are its cloud bands. The outstanding features of Saturn are its incomparable rings. In addition to these sights and much detail that can be glimpsed with larger telescopes or by experienced observers, Jupiter has four moons easily bright enough to see in binoculars. Saturn has even more, though they are fainter.

BELTS AND ZONES OF JUPITER

The illuminated surface area that Jupiter presents in the telescope is greater than that of all the other planets combined. Even a rather small telescope can make Jupiter look as big as the Full Moon does to the naked eye, and show some of the patterns in this kingly planet's clouds.

Find yellow-white Jupiter in your telescope. If the atmosphere is reasonably steady—if the "seeing" is good—then your telescope almost certainly shows at least one or two darkish bands running across the globe. These bands run parallel to each other; the ones you are most likely to see are the equatorial ones, with a light strip in between them. Now, with more concentration, or a bigger telescope, you begin detecting other parallel dark bands—shorter ones, obviously shorter because they pass west-east across Jupiter at higher latitudes. And then detail within and between the bands begins to glimmer briefly into sight.

The dark bands on Jupiter are known as **belts**, and the light bands between them as **zones**. They run like strips of latitude around Jupiter, the result in part of the huge planet's incredibly fast spin. (Jupiter's "day" is less than ten hours long, even though the planet is eleven times wider than Earth.) A better look at the belts reveals color—mostly brownish red—and structure within them. Among Jupiter's other atmospheric features, the most

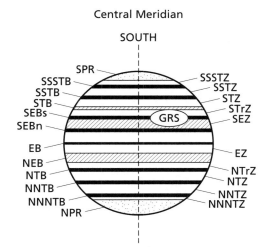

Figure 31.
Jupiter's belts and zones.
(GRS is the Great Red Spot.)

famous is the *Great Red Spot (GRS)*, a kind of oval-shaped storm much larger than the Earth that has persisted for at least a few hundred years. The GRS is usually not very intensely colored, and it's important to try to observe it when it is near Jupiter's central meridian. Predicted times for the GSR's passage across the central meridian are often listed in or near the "Calendar Notes" section of *Sky & Telescope* magazine (see Sources of Information).

GALILEAN SATELLITES OF JUPITER

The four largest moons of Jupiter were discovered by Galileo when he turned his first tiny telescope on the planet. You can glimpse a few of them with steadily held or mounted binoculars on some nights. In fact, several of them have been glimpsed with the naked eye under good conditions. They are bright enough, but the planet's nearby light tends to overwhelm them for the naked-eye observer. A telescope is certainly best for enjoying the full show of these four worlds.

Floating beside Jupiter in your telescope are one, two, three, or four little stars, often more or less in line. Sometimes one is near the planet or near another moon and you can tell that the distance between them is changing in a matter of minutes. If one is missing, an hour or two later you may see it, suddenly emerged from behind or in front of the planet—or from Jupiter's shadow. If you look tomorrow night, you will find an entirely different arrangement of these little jewels. Are they merely points of light? With a good medium-size telescope you will see that these "stars" have a little fatness. And under very good conditions these tiny disks start becoming visible at a high magnification (say 300× or more).

The Galilean satellites are, in order from innermost to outermost, Io, Europa, Ganymede, and Callisto. What our spacecraft have taught us about them is nothing short of stunning. Io is a world of erupting volcanos. Europa has a smooth but cracked ice surface under which an ocean of water may exist. Ganymede and Callisto are, respectively, the largest and third largest moons in the solar system, and the former is now known to have a thin atmosphere, a major magnetic field, and even an aurora of sorts.

With a telescope, an observer can eventually see all manner of events involving these moons and Jupiter. When one of the moons passes behind Jupiter, the event is called an occultation. When one of the moons goes through Jupiter's shadow, it is an eclipse. When one of the moons passes in front of Jupiter, it is a **transit**. If the shadow of a moon crosses in front of Jupiter (or, rather, appears on the cloud tops of Jupiter), the event is called a **shadow transit**. Transits are usually harder to see than shadow transits. When Jupiter lies near quadrature, forming a right angle with Sun and Earth, the transiting moon and its shadow are at their most separated, and, more importantly, Jupiter's shadow can be projected well to one side of the planet's disk from our point of view so that eclipses can take place long before or after the moon goes behind the planet in an occultation.

Montage of Jupiter and its largest moons. (NASA photographs)

THE RINGS OF SATURN

The rings of Saturn can often be detected even in quite small telescopes. Their exquisite perfection sends shivers down an observer's spine more than any other first sight in astronomy, except perhaps the first telescopic view of the Moon. The rings are composed of countless icy moonlets ranging in size from mountains and boulders to tiny specks. The span of the visible rings is more than 170,000 miles.

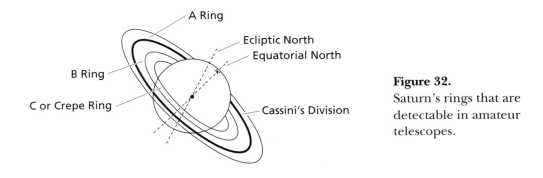

Figure 32. Saturn's rings that are detectable in amateur telescopes.

The first thing to notice—after you've finally caught your breath—is that the rings are indeed plural. Even small telescopes can distinguish between the inner B ring that is brighter, whiter, and broader than the outer A ring. What separates these two rings?

As you admire the rings of Saturn in your telescope, you glimpse a thin dark line dividing them, inner from outer. You may need a six-inch or larger telescope and good "seeing" to be doing this, but once you do, you can say that you have beheld a gap about as wide as the Atlantic Ocean—but out at a distance of nearly a billion miles! You are seeing Cassini's Division.

Other gaps between and within Saturn's rings are sometimes visible from Earth. But you might be more likely to see, between the B ring and the planet, the elusive C or crepe ring. It may look like nothing more than a gauzy, hazy glow. At some times its visibility is greater than at others.

The most important long-term factor affecting the overall visibility of the rings is the degree to which they are tilted toward Earth. In 1995–96, the rings went through a series of edgewise presentations to Earth and the Sun. The rings were sometimes darkened; sometimes they appeared no thicker than a line of light, or even so thin when presented perfectly edge-on that they were invisible. The performance was strange and captivating but not always easy to observe. Since then, Saturn's position in its orbit has continued to present the rings at more and more of an angle to Earth. They will reach a maximum of about 27 degrees in 2002. The more tilted or "open" the rings are, the more spectacularly bright they are and the better observable most of their features. So the period from about 1998 to 2006 is quite favorable for looking at Saturn's rings. This is also true because the planet reaches the perihelion point of its almost thirty-year-long orbit in 2003. Then the planet, both rings and ball, will be at its biggest and brightest to an observer on Earth.

THE GLOBE AND MOONS OF SATURN

The ball and satellites of Saturn are often overlooked by observers who are mesmerized by the rings. Saturn's globe is the most oblate—the most flattened in the pole-to-pole diameter—of any planet. It possesses dark belts and light zones like Jupiter's, only they are much less prominent. Often, however, it is possible to see the shadow of the rings as a dark line running across the planet, or the shadow of the planet on the rings where they run behind the globe.

The satellite system of Saturn is a throng. As many as six moons of Saturn are within range of medium-size telescopes, though you might have trouble distinguishing some of them from background stars. The brightest Saturnian moon—about eighth magnitude— is the aptly named giant Titan. Titan has an atmosphere denser than Earth's and is eternally shrouded in an orange haze. With a large amateur telescope you might notice a hint of that color in the starlike point of Titan.

SUMMARY

Jupiter and Saturn are the most dependable of planets for observers, visible for about eleven months out of the year and always offering an interesting sight in telescopes. Jupiter displays easily visible dark belts and light zones in its clouds, but its famous Great Red Spot is usually much more difficult to detect. Jupiter's four big moons—Io, Europa, Ganymede, and Callisto—would be regularly visible to the naked eye if they weren't so close to the overwhelmingly brilliant Jupiter. Through the telescope they offer endless entertainment with their ever-changing patterns and their occultations and eclipses by Jupiter, their transits and shadow transits of Jupiter. The rings of Saturn are thrillingly visible even in small telescopes. A slightly larger telescope can distinguish between the broader, brighter B ring and the outer A ring, and can detect the dark line of Cassini's Division between them. The gauzy C or crepe ring is much more elusive. Up to six of Saturn's moons can be seen in medium-size telescopes, the brightest being Titan.

Montage of Saturn and its largest moons. (NASA photographs)

THE PHASES OF VENUS
AND MERCURY

✦   ✦   ✦

Time: Within a few hours after sunset or before sunrise, whenever

Venus or Mercury is far enough from the Sun in the sky.

GALILEO'S DETECTION of the phases of Venus with his telescope was of tremendous importance. Some astronomers, such as the famous Tycho Brahe, had theorized that Mercury and Venus did orbit the Sun, but that the Sun in turn orbited the Earth. But the kinds of phases Galileo saw Venus present were only consistent with Venus and the Earth orbiting the Sun. This observation provided important support for Copernicus's heliocentric (sun-centered) system.

For today's observers, the changing phases of Venus and Mercury are a swift and fascinating show to watch. It's a good thing, too, because neither planet shows much detail in its clouds (Venus) or surface (Mercury) as viewed from Earth.

THE PHASES OF VENUS

*Y*ou *are looking at something shaped just like a crescent Moon in your telescope. But unlike the dull yellow of the Moon, this crescent is intensely bright, even before the sky behind it gets dark. A half-hour later you get brilliant Venus back in your telescope and what do you see? With an almost fully dark sky, the crescent is truly dazzling; you have some trouble defining its edges, let alone seeing any hint of detail on it. But is one of the crescent's tips distinctly longer than the other?*

This imagined observing session is just one, and far from the most exciting, that you may have with Venus. It's true that as an evening apparition begins, with Venus moving away from superior conjunction, the planet looks small (about twelve seconds of arc wide—about one-third Jupiter's minimum width and a quarter of Jupiter's maximum width). And you may have trouble seeing that this globe of Venus is not quite fully lit—it is only a little bit gibbous. But as Venus comes to greatest evening elongation, it is growing in length and shrinking in illuminated fraction—down to a striking half-Venus.

Theoretically, Venus should be exactly half-lit at greatest elongation. In practice, however, the half-lit presentation—called **dichotomy**—seems to happen a few days before greatest evening elongation and a few days after greatest morning elongation. The discrepancy probably arises from the fact that Venus is not a perfectly smooth ball but rather a world enshrouded in clouds that do have some structure and thickness.

To judge from what they look like in the telescope, you might think the clouds of Venus are completely featureless. Then you start seeing streaks of shadowy darkness here and there on the bright surface, and deformations of the crescent, some of which subsequently prove to be optical effects from your telescope. But some of these elusive markings result from the cloud structure.

For Venus watchers, it is the months from evening greatest elongation to inferior conjunction and back out to morning greatest elongation that are electrifying. Observers sometimes think they see a pale glow on the night part of Venus—the still unexplained "ashen light." The crescent becomes breathtakingly skinny and amazingly long; the crescent has been glimpsed with the naked eye occasionally when Venus is a few weeks from inferior conjunction and up to one minute of arc long. It can certainly be glimpsed in steadily held or mounted binoculars at such times. Very close to inferior conjunction, the crescent's ends have been seen to extend and form a full circle of light—light shining through the atmosphere of Venus.

OBSERVING MERCURY

For beginners, getting a good naked-eye view of Mercury should be a thrill. According to one legend, the great Copernicus never got to see the planet in his entire life. But if you have a telescope handy, try for the additional thrill of seeing the globe of this speedy little sun-baked world.

It's essential that you view Mercury as high in the sky as possible. The effects of bad "seeing" low in the sky almost always make the image shaky and blurry. But by finding Mercury early in dusk or following it until late in dawn near a favorable greatest elongation, you will eventually succeed in getting a better view. The good view of the then approximately half-lit Mercury is a rare treat. Often slightly orange or pink from haze, Mercury is capable of showing a hint or two of dark markings on its form, but you shouldn't count on making such a detailed sighting tonight—or this year.

TRANSITS

Talk about rare: that's the word for a marvelous kind of event that we can see only Mercury or Venus perform. I'm referring to a transit across the face of the Sun.

As we've seen, solar and lunar eclipses don't occur at every New Moon and Full Moon because the Moon usually passes slightly north or south of the line through Sun and Earth. Transits don't occur at every inferior conjunction of Mercury and Venus with the Sun because the planets almost always pass north or south of the Sun as seen from Earth.

I can tell you about my experiences of seeing the little but intensely black dot of Mercury crossing my projected image of the Sun on two different mornings in the early 1970s. Since then, there have been no opportunities to see a transit of Mercury from the eastern United States, though there have been some chances from elsewhere on Earth—at the rate of about two a decade.

What I can't tell you about is the experience of seeing a transit of Venus. No one alive today can: the most recent one occurred in 1882! Fortunately, these incredibly rare events occur in pairs, and the next pair of transits of Venus is in our relatively near future: June 8, 2004, and June 6, 2012. The end of the 2004 transit will be visible at sunrise in eastern North America.

SUMMARY

The phases of Venus go from nearly full and small apparent size just after superior conjunction, to about half-lit ("dichotomy") close to the time of greatest elongation, and to a skinny but very long crescent around the time of inferior conjunction. Many observers have seen a dim illumination of the night side of Venus called the "ashen light." Subtle dark markings can sometimes be seen in the clouds of Venus. Mercury must be seen as high as possible to get a reasonably steady image of it, which means it is usually seen at its best when about half-lit near greatest elongation. Transits of Mercury and Venus are rare events when the dark shapes of these planets can be viewed crossing in front of the Sun's disk (using the proper safe solar viewing techniques discussed earlier in this book).

✦ NIGHT 34 ✦

FAR MARS AND NEAR MARS

✦ ✦ ✦

Time: Whenever Mars is visible.

FOR OBSERVERS, MARS IS ALMOST two different planets. One Mars is tiny—truly a speck, little bigger than a star in small telescopes. It is outshined by most of the first magnitude stars. This, unfortunately, is the Mars we get for the better part—actually the worst part—of two years. But then there is the other Mars. This Mars is as big as Saturn or bigger in the telescope with an exquisite, chillingly white polar ice cap; on some nights it brims with the intricate detail of dark markings and white and gold touches of frosts or fogs and dust storms. This Mars is as bright as Sirius or even much brighter, its fiery orange-gold imposingly prominent.

We only get the brighter Mars for a season or two in each period of more than two years. But, as explained earlier, there is a much longer cycle in which each of the oppositions of Mars is poorer than the last until a "least good" one, and then steadily better until a "perihelic opposition" in which Mars often outshines even Jupiter and dominates the sky like a blazing ember.

APHELIC AND PERIHELIC OPPOSITIONS

Figure 33 shows the progress of Mars from its last perihelic opposition in 1988 to the aphelic opposition of 1995 and on to the very close perihelic opposition of 2003. Table 5 gives observing information for these oppositions and reveals some remarkable points. For instance, at the aphelic opposition of 1995 the apparent angular diameter of Mars was about 14 arc-seconds, but at the perihelic opposition of 2003 it will be 25 arc-seconds. That's a substantial difference in size, but the difference in surface detail visible on Mars is far greater still. An important threshold exists beyond which the number of detectable features begins to multiply.

On the other hand, look at the difference in maximum altitudes of Mars between the 1995 and 2003 oppositions. Mars was exactly twice as high at culmination (crossing of the north-south meridian) at the 1995 opposition as it will be in 2003, as seen from latitude 40° N. Mars will be higher in 2003 as seen from the middle latitudes of the southern hemisphere.

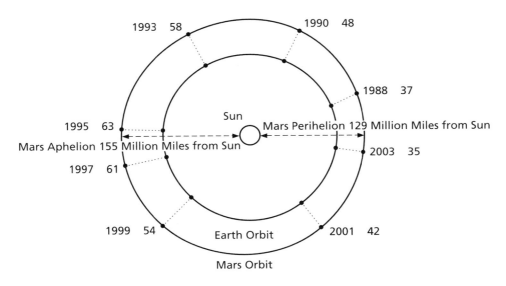

Figure 33.
Mars's oppositions, 1988–2003. Numbers next to the years
are Earth–Mars distances in millions of miles.

One other factor determines how well we will see certain features on Mars: the tilt of Mars to us and to the Sun around the time of opposition. The 1997 opposition of Mars occurred with the north polar cap of the Red Planet presented to us at almost the maximum tilt possible. That's very favorable—except the fact that the northern hemisphere of Mars was only four days past its summer solstice at the time of opposition meant the pole was near minimum extent (and possibly hidden by a hood of cloud). Every opposition of Mars has both favorable and unfavorable aspects for observation; every one is worth watching.

WATCHING A GROWING MARS

*C*onsider the progress of Mars from its farthest to its nearest point in one synodic period of about two *years and two months. What stage is Mars currently in? Use the descriptions below to see what you can on the planet.*

When I wrote above that Mars is a speck in telescopes when it is on the opposite side of its orbit from us, I meant it. At these times Mars is actually smaller in apparent size than Uranus, which lies almost two billion miles away. Today's bigger amateur telescopes and electronic imaging have made even this tiny a Mars not entirely a lost cause. But for most amateur astronomers, those who possess medium-size or small telescopes, the long periods when Mars is this size are times when viewing it is little more than a curiosity: they won't see any detail on the planet.

TABLE 5

Observing Information for Mars's Oppositions

Opposition	Magnitude	Altitudes[a]	Nearest Earth	Distance[b]	Size[c]
1988 Sep 27	−2.7	48°	1988 Sep 21	37	24
1990 Nov 27	−2.0	73°	1990 Nov 19	48	18
1993 Jan 7	−1.4	76°	1993 Jan 3	58	15
1995 Feb 11	−1.2	68°	1995 Feb 11	63	14
1997 Mar 17	−1.3	55°	1997 Mar 20	61	14
1999 Apr 24	−1.7	38°	1999 May 1	54	16
2001 Jun 13	−2.3	23°	2001 Jun 21	42	21
2003 Aug 28	−2.9	34°	2003 Aug 27	35	25
2005 Nov 7	−2.3	66°	2005 Oct 29	43	20

[a] Angular altitude at highest (middle of night) as seen from 40°N latitude.
[b] Distance of Mars from Earth in millions of miles.
[c] Maximum disk diameter of Mars in arc-seconds.

Then comes the stirring. Mars grows to 5 arc-seconds wide, 7 arc-seconds wide; a polar ice cap may start to become visible. Mars is seen rising at late-night bedtimes. When its magnitude reaches 0, its angular diameter reaches 10 arc-seconds. The period during which the modestly equipped amateur astronomer can make useful (and thrilling!) observations of Mars has begun.

It is not easy to see detail on a 10-arc-second-wide Mars. Your telescope must have its optical components in proper alignment and preferably be larger than three or four inches in aperture, and Mars must be reasonably high on a night of good "seeing." Even then, the observer must learn to wait patiently for those moments of really superior "seeing" and be able to grab and remember the details that leap out briefly.

Even when Mars is 15 arc-seconds across, it can be a bust if not observed skillfully. As for 20 arc-seconds, 25 arc-seconds—well, then the question is whether you just see some features or, with knowledge and skill, perceive the intricate gray and green patchworks and the dozens of details of this neighbor world.

FEATURES ON THE MOST EARTHLIKE WORLD

Mars is startlingly different from Earth, but it also has an amazing number of similarities. The axis of Mars is tilted almost exactly as much as ours, so its seasons are comparable. Its

Mars mosaic.
(NASA photographs,
processed by USGS)

day is only a bit longer than ours. And Mars sports polar ice caps that grow in winter and shrink in summer, just like Earth's.

But what makes Mars potentially most like Earth and most exciting are some things that we find only traces of. Dry riverbeds and floodplains indicate that large quantities of water once flowed on Mars. The undulations in the frozen mud around some Martian craters are evidence that the energy of the original meteor impacts released lots of water frozen in the soil or below it. In 1996, it was announced that scientists may have found 3½ billion-year-old fossils of past Martian life—something like bacteria—within a meteorite believed to have been blasted from Mars to Earth. In 1997, *Pathfinder* and its mini-rover *Sojourner* found evidence of multiple floods in the distant past at their landing site.

Are there fossils on Mars? Could Martian life exist somewhere beneath its surface now? What will the space probes of the late twentieth and early twenty-first century tell us about this neighbor world and its mysteries? How soon will human beings visit Mars in person and gaze up at its pink skies and blue sunsets and its two tiny, craggy, hurtling moons? These are among the questions that will haunt your imagination as you stare at Mars in the telescope.

But what do you actually see? Most of the globe appears orange, the color of the sands on this frigid yet desert wilderness world. Under good conditions, however, you may see not just a polar ice cap but the dark, seemingly grayish green markings on the planet. It was once thought that these might be areas of vegetation, for they appeared to expand in the hemisphere where the ice cap was melting as spring came. In reality, the greenish color is an effect of contrast with the orange sands, and its seeming darkening is really a brightening of the orange areas as frosts melt off and expose them more clearly. Changes in the shapes of the dark markings can be attributed to the dust storms of Mars blowing the sands around.

Despite these seasonal changes, many of the dark areas survive and maintain approximately the same shape and prominence as the decades and centuries pass. Most prominent of these regions is the roughly triangle-shaped Syrtis Major, first noticed by Dutch scientist Christiaan Huygens as early as the middle of the seventeenth century. Figure 34 gives the approximate appearance of other major dark regions of Mars. The prominence of some of these regions depends much on recent dust storm activity. In particular, Solus Lacus is only obvious enough at some oppositions to earn its nickname, "the Eye of Mars."

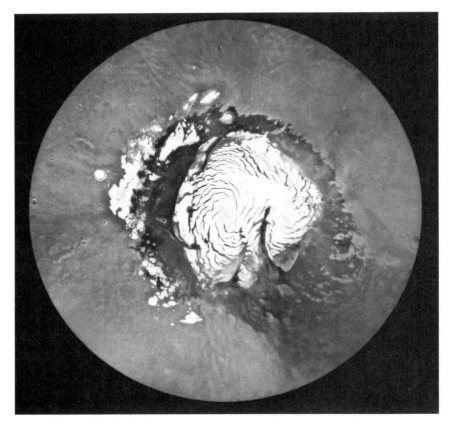

Martian polar ice cap. (NASA photograph, processed by USGS)

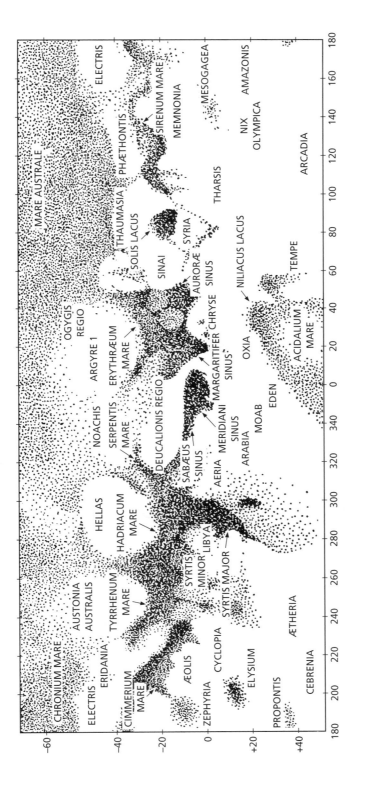

Figure 34.

Map of Martian surface features as they appeared at the close opposition of 1986. (Drawn by Doug Myers)

What a challenge it is to perceive many of these dark markings—a wonderful challenge. And the best way to sharpen your ability to see these details is to try sketching them.

You can learn what features will be on the central meridian on Mars at different times and dates around opposition by looking up the information in one of the popular astronomy magazines (see Sources of Information). But actually, once you have identified a very noticeable Martian surface feature, you should be able to keep track of which longitude of Mars you're looking at for many nights or even weeks. The reason is that any feature you see on the central meridian of Mars at a certain time one night will reach the central meridian about thirty-six minutes later each night. This is simply a consequence of Mars taking that much longer than Earth to make one full rotation. This relationship is quite convenient in some ways, but it also means that once a feature of Mars starts coming to your meridian too late for you to see (after Mars sets or daylight comes), it won't be back in view from your longitude for more than a month. So every week of the few months around a Martian opposition is important for observers.

SUMMARY

Mars appears as a tiny, featureless dot in telescopes for many months around its time of conjunction with the Sun. But as it approaches opposition, it can swell to anywhere from about five to eight times wider and begin to display dozens of intricate dark markings on its ochre globe. There is a great difference between what can be seen at an aphelic versus a perihelic opposition of Mars; the latter are very favorable events that take place only at intervals of fifteen or seventeen years. The dark markings on Mars are regions of rock or soil darker than the surrounding orange sands. Syrtis Major is the most prominent of them. The same markings arrive on the central meridian of Mars about thirty-six minutes later each night.

✦ **NIGHT 35** ✦

A Variety of Double and Variable Stars

✦ ✦ ✦

Time: Any night.

NAKED-EYE OBSERVERS can detect a few double stars. A few variable stars can be followed through their entire range of brightness with the naked eye, and many more can be glimpsed at peak brightness.

But unaided vision can show us only a tiny sampling of the rich beauty of double and variable stars that are available to observers using binoculars and telescopes. In addition, some aspects of these stars can't be seen with the naked eye. For instance, there are no double stars bright enough for the naked eye to perceive the color of both members of the pair. But through the telescope the contrasting tints of many doubles is their chief beauty. Likewise, even the brightest of (fairly regular) long-period variables, Mira, has a far dimmer minimum than the naked eye can see. How can you fully appreciate its huge magnitude range if for half of that range you can't view it at all?

Get out a telescope. Welcome to the variety and full set of beauties to be found from double stars and variable stars.

Many Kinds of Double Stars

Although the term "double star" is often used to mean a star system with more than one star, we also refer to "multiple star" systems featuring three, four, or even more stars! As a matter of fact, the star in the brightest part of the Great Nebula in Orion splits into four as seen in small telescopes, a quartet commonly known as the Trapezium—but two more stars of the system can be glimpsed in larger amateur telescopes, and another two remain beyond such telescopes' grasp. Another plentifully multiple star is Gemini's almost first magnitude Castor, though some of the members cannot be detected visually. Members of a star system that are detected by their effect on the overall spectrum are **spectroscopic**

binaries; those detected by their effect on the system's long-term motion through space are **astrometric binaries**.

Multiple stars are impressive because of their number, but perhaps nothing is so beautiful as contrasting colors in two or more stars in a system. In most cases the colors would not be as vivid by themselves—but their contrast with each other accentuates the hues. One of the finest combinations is a reddish or orange star with a bluish or greenish one. Two famous examples of double star systems with this color combination are Albireo (Beta Cygni) and Almak (Gamma Andromedae).

Usually one star is brighter than the other, but sometimes it is remarkable when the stars are of similar brightness and color. Porrima (Gamma Virginis) consists of two essentially identical stars, which have been described as looking like car headlights heading for us. Kuma (Nu Draconis) is a dimmer but also equal pair—and very widely separated, especially as compared with the currently converging and already tight pair of Porrima. Epsilon Lyrae is the famous "Double Double" near Vega. Very sharp-eyed people can sometimes make the first split, into two similarly bright stars, and binoculars certainly can do it. But it usually takes 100× magnification or more to split each of these stars into further pairs of similarly bright stars.

The separation of double stars is a key factor in their visibility, of course. Many are easy to split even with very low magnification. Others are so close together that they provide a supreme test of telescope optics and atmospheric "seeing." The larger the aperture—the

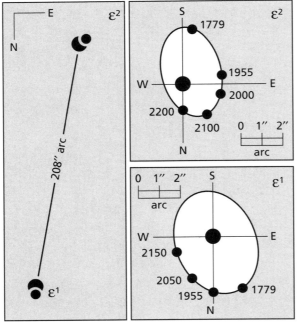

Figure 35.
Epsilon Lyrae diagram.

diameter of the primary mirror or lens—of your telescope, the greater resolving ability it has with which to show fine detail on the Moon and planets and to split close double stars. How large an aperture do you need to split various double stars? Table 6 lists approximate values in arc-seconds for the separation that telescopic apertures of various sizes can achieve with reasonably bright stars of identical brightness on nights of the best "seeing." The greater the contrast in brightness between two component stars, the more difficult it will be to see the fainter one at a given close separation.

There is a pleasure in seeing double star components of similar brightness, or in splitting vexingly close double stars. But it can also be gratifying to observe star pairs that are of drastically different brightness. Vega and Polaris are two examples of famous bright stars with tremendously dimmer suns near them that can be detected in medium-size amateur telescopes. Suddenly, just beside the rays of the bright star in your telescope, you catch sight of a tiny but intense speck of precise brightness: the companion star. In both these cases, though, the little stars are believed to be just chance line-of-sight objects (optical doubles), not actually sharing motion in space with Vega or Polaris.

A classic example of an optical double, in which there is somewhat less difference in brightness between the two component stars, is Delta Herculis, also known as Sarin. But these components are of magnitude 3.1 and 8.2—the latter about a hundred times dimmer than the former. An intriguing speculation of recent years is that the famous gold and blue Albireo could be an optical double.

Double stars are everywhere. It is believed that more than half of all stars belong to double or multiple systems. Even our own Sun may yet be found to have a dim companion many times farther out than Pluto. *Use Tables 6 and 7 as a starter list to observe some of the most beautiful of these systems.*

TABLE **6**

Double Star Separations and Telescope Apertures

Aperture	Arc-Seconds
1-inch telescope	4.56″
2-inch	2.28″
3-inch	1.52″
4-inch	1.14″
6-inch	0.76″
8-inch	0.57″
10-inch	0.46″
12½-inch	0.36″
16-inch	0.29″

TABLE 7

Selected Double Stars

Star	RA h	RA m	Dec. °	Dec. ′	Mag.	PA[a] °	Dist. ″
Gamma Arietis	01	53.5	+19	18	4.8, 4.8	0	7.8
Gamma Andromedae	02	03.9	+42	20	2.3, 5.1	63	9.8
Beta Orionis[b]	05	14.5	−08	12	0.1, 6.8	202	9.5
Eta Orionis	05	24.5	−02	24	3.8, 4.8	80	1.5
Theta Orionis[b]	05	35.4	−05	25	4.9, 5.0	314	135
Theta-1 AB					6.7, 7.9	—	8.8
Theta-1 AC					6.7, 5.1	—	12.8
Theta-1 AD					6.7, 6.7	—	21.5
Theta-2					5.2, 6.5	92	52
Zeta Orionis	05	40.8	−01	57	1.9, 4.0	165	2.3
Beta Monocerotis	06	28.8	−07	02	4.7, 4.8	132	7.3
Alpha Geminorum[b]	07	34.6	+31	53	1.9, 2.9	65	3.9
k Puppis	07	38.8	−26	48	4.5, 4.7	318	9.9
Zeta Cancri	08	12.2	+17	39	5.1, 6.2	72	6.0
Gamma Leonis	10	20.0	+19	51	2.2, 3.5	125	4.4
Xi Ursae Majoris	11	18.2	+31	32	4.3, 4.8	273	1.8
Alpha C. Venaticorum	12	56.0	+38	19	2.9, 5.5	229	19.4
Zeta + 80 Ursae Majoris[b]	13	23.9	+54	56	2.3, 4.0	71	709
Zeta Ursae Majoris	13	23.9	+54	56	2.3, 4.0	152	14.4
Epsilon Bootis	14	45.0	+27	04	2.5, 4.9	339	2.8

A VARIETY OF VARIABLE STARS

Earlier in this book we examined Algol (Beta Persei), Beta Lyrae, and Delta Cephei as examples of naked-eye variable stars whose brightness changes are easy to study on a weekly basis. But there are thousands of variable stars that amateur astronomers regularly monitor; millions of observations of them have been officially recorded.

Eclipsing binaries are the class to which Algol belongs. One star eclipses the other, or both each other, with the pair being so close that visual separation of them is typically impossible. But we can figure out what must be happening (often with the help of examining the spectrum of the system). When one star passes in front of another, the total light coming to us from the system is diminished—especially when it is the brighter star being covered by the dimmer, as seen from our point of view. The eclipses may be total or they may be partial—the latter being the case with Algol.

Star	RA h	RA m	Dec. °	Dec. ′	Mag.	PA[a] °	Dist. ″
Alpha Librae	14	50.9	−16	02	2.8, 5.2	314	231
Mu Bootis	15	24.5	+37	23	4.3, 6.5	171	108.3
Beta Scorpii	16	05.4	−19	48	2.6, 4.9	21	13.6
Nu Scorpii	16	12.0	−19	28	4.2, 6.1	337	41.1
Alpha Scorpii[b]	16	29.4	−26	26	var., 5.4	273	2.6
Zeta Scorpii	16	54.3	−42	20	3.6, 4.8	—	408
Alpha Herculis	17	14.6	+14	23	var., 5.4	104	4.6
36 Ophiuchi	17	15.4	−26	33	5.1, 5.1	152	4.7
Nu Draconis	17	32.2	+55	11	4.9, 4.9	312	61.9
Epsilon Lyrae[b]	18	44.3	+39	40	4.7, 4.6	173	207.7
Epsilon-1					5.0, 6.1	350	2.6
Epsilon-2					5.2, 5.5	82	2.3
Theta Serpentis	18	56.2	+04	12	4.5, 5.4	104	22.3
Beta Cygni[b]	19	30.7	+27	58	3.1, 5.1	54	34.4
Alpha Capricorni	20	18.1	−12	33	3.6, 4.2	291	378
Beta Capricorni	20	21.0	−14	47	3.4, 6.2	267	205
Gamma Delphini	20	46.7	+16	07	4.5, 5.5	268	9.6
Xi Cephei	22	03.8	+64	38	4.4, 6.5	274	8.2

Note: The double stars listed are chosen from among those visible from 40°.

[a] PA is position angle: 0° = north, 90° = east, and so on back around to north.

[b] Beta Orionis is Rigel; Theta-1 Orionis is Trapezium; Alpha Geminorum is Castor; Zeta Ursae Majoris and 80 Ursae Majoris are Mizar and Alcor; Alpha Scorpii is Antares; Epsilon Lyrae is "the Double Double"; Beta Cygni is Albireo.

Cepheids are the wonderful "standard candles" used for measuring distances in space. This is possible because the length of a Cepheid's period from maximum to maximum is proportional to its luminosity—which, once known, can enable us to compare it to the star's apparent brightness and compute its distance. Delta Cephei itself is the classic example, but many others abound. In these stars, actual changes in size and luminosity (pulsations) occur. The changes take place over a period of remarkably precise and unchanging duration.

Long-period variables are generally "red giant" stars whose vastly distended outer atmosphere of gases expands and contracts over periods of months or years, either semiregularly or irregularly. The most famous example of such a star is Omicron Ceti, better known as Mira—a name which means "the wonderful." Mira typically goes from a minimum of eighth magnitude or dimmer up to a maximum of third magnitude, or occasionally, brighter. Far too faint to be seen with the naked eye at minimum, this star in the neck of Cetus the Whale can brighten to rival or even outshine magnitude 2.5 Alpha Ceti in the

Whale's head. In the nineteenth century, Mira once became almost as bright as first-magnitude Aldebaran! Its period is about eleven months, so it tends to be at its best about a month earlier each year. This means that for a few years in each dozen, this star reaches maximum when it is too near to conjunction with the Sun to be observed properly. The very bright peak in 1997 occurred in February, so the star should be well-placed at maximum for the final years of the twentieth and first years of the twenty-first century.

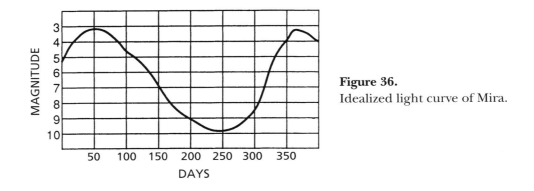

Figure 36.
Idealized light curve of Mira.

The most spectacular of variable stars are the **novae** and **supernovae**. Both kinds of stars suffer tremendous losses of material in cataclysmic events.

A nova usually involves a small fraction of a star's mass being thrown off, causing the star to become maybe tens of thousands of times brighter for a few weeks before slowly fading toward its original dimness (or becoming even fainter) over a period of months. There do seem to be stars that are like slightly less extreme versions of novae whose outbursts recur within a hundred years or less. The most famous of these is T Coronae Borealis—"the Blaze Star"—which on two known occasions has flared up mightily to second magnitude before quickly fading.

A supernova involves a star imploding—its outer layers collapsing on its core—and then exploding. In one of the most violent acts in the universe, the star ejects much of its material at speeds of millions of miles an hour and radiates as much light as billions of ordinary stars. A supernova can outshine the light of all the billions of stars of its galaxy combined. The surviving star can end up as a **neutron star** (of which one particular kind is a **pulsar**) or as a **black hole**. These exotic stars are so densely packed with matter that a thimbleful of one would weigh from thousands to millions of tons.

It's impossible to predict when the next nova will appear in our skies, but a number of amateurs spend much time looking for—and in many cases have discovered—a nova. Most novae have been located near the Milky Way band in the sky.

TABLE 8
Selected Variable Stars

Name	Type	Magnitude Range[a]	Period (days)	RA	Declination
Mira	long-period	1.7–10.1	332	2h 19.3m	−2° 59′
Rho Persei	semiregular	33–55	33–55	3 05.2	+38 50
Algol	eclipsing	2.1–3.4	2.87	3 08.2	+40 57
Betelgeuse	semiregular	0.0–1.3	2335[b]	5 55.2	+7 24
Zeta Geminorum	Cepheid	3.6–4.2	10.15	7 04.1	+20 34
U Hydrae	irregular	4.3–6.5	450?	10 37.6	−13 23
R Hydrae	long-period	3.5–10.9	389	13 29.7	−23 17
R Coronae Borealis	irregular	5.7–14.8	—	15 48.6	+28 09
Alpha Herculis	irregular	2.7–4.0	—	17 14.6	+14 23
Beta Lyrae	eclipsing	3.3–4.4	12.94	18 50.1	+33 22
Chi Cygni	long-period	3.3–14.2	408	19 50.6	+32 55
Eta Aquilae	Cepheid	3.5–4.4	7.18	19 52.5	+01 00
Mu Cephei	irregular	3.4–5.1	730[c]	21 43.5	+58 47
Delta Cephei	Cepheid	3.5–4.4	5.37	22 29.2	+58 25
Beta Pegasi	irregular	2.3–2.7	—	23 03.8	+28 05
Rho Cassiopeiae	irregular	4.4–5.2	—	23 54.4	+57 30

[a] "Magnitude Range" given may be more extreme than the mean range in the case of some of the long-period, semiregular, and irregular variable stars.
[b] Also shorter subperiods of 200–400 days.
[c] Secondary period of 4,400 days!

The last supernovae in our galaxy that were properly observed occurred about four hundred years ago—just before the invention of the telescope. Are we overdue for another one? Perhaps, though no one can say for sure. Of course, the chance that you or I will be its discoverer is remote. But that won't matter if we live to see it; seeing the next supernova in our galaxy will be reward enough. The last one flared up to become as bright as Jupiter; the one before that was brighter than Venus. And the one in the year 1006 rivaled the half-Moon in brilliance!

We may not be seeing a supernova soon, but the accompanying table lists other examples of variable stars for you to monitor.

SUMMARY

Double star systems can contain more than two stars and be "multiple stars." The colors of the two members of a double star system can be accentuated if they contrast with each other. The larger the aperture of a telescope, the closer the separation of a double star that it can split. It is easier to split component stars of similar brightness. Doubles too close together to be seen or photographed separately are sometimes recognized by studying the point of light's spectrum or long-term motion.

There are many types of variable stars. Some of the most important classes are eclipsing binaries, Cepheids, long-period variables, novae, and supernovae. A new nova occurs every few years (or more often), and many have been discovered by amateur astronomers. A few stars seem to be recurrent novae. Supernovae are far greater explosions and brightenings of stars than novae, ejecting more of the star's mass and producing neutron stars and black holes.

✦ NIGHT 36 ✦

STAR CLUSTERS AND NEBULAE

✦   ✦   ✦

Time: Any night.

WE FOUND EARLIER in this book that there were only a few star clusters that appear to the naked eye as groupings of stars rather than just fuzzy spots of light. But with telescopes there are thousands of star clusters that can be resolved entirely or partly into their individual components—an endless assortment of stellar jewelry.

We also found earlier in this book that only a few nebulae can be detected with the naked eye. But the telescope brings dozens of them into potential sight.

Your observational quest for this Night is to observe as many of the clusters, nebulae, and other objects in the accompanying tables as possible.

CLUSTERS, ASSOCIATIONS, MOVING GROUPS

Back on Night 23, we distinguished between open (also called galactic) star clusters and globular star clusters. But the naked eye can reveal few globular clusters and certainly cannot show them to be what they are: the immense congregations of hundreds of thousands

Omega Centauri. (Photographed by Akira Fujii)

of stars. The naked eye cannot distinguish between a distant open cluster and a globular. But the view through a telescope is a different matter.

Whereas only the Pleiades appear as a really rich concentration of stars to the naked eye, most bright open clusters resolve into large numbers of individual stars in a telescope. Even the incredibly distant globular clusters can begin to be resolved into their components in a medium-size or large amateur telescope. The globular's outlying stars look like pinpricks of light glittering here and there through the cloud-like glow of the more central, blended together stars.

Are there arrangements of stars looser than open star clusters that are going through space together? Yes, a scattering of young stars that were born together (in a nebula) may not yet have flown enough apart from each other for us to miss their kinship. Such a **star association**, composed of young, very hot stars, can be found in Perseus and includes the star Zeta Persei.

A selection of older stars, aged into different spectral types and scattered over a rather large region of the sky, may nevertheless be found to have similar motions. An example of such a **moving group** may be the one that includes five of the stars of the Big Dipper (all but the ones at either end) and a number of other famous stars spread over much of the winter and spring sky. This is more often now called the Ursa Major Cluster. But the concept of it being a "moving group," with less certain outermost members forming part of an

Ursa Major Stream, remains useful. Incidentally, one of the stars of the Ursa Major Stream could be Sirius. And if all of these stars are truly going through space together, then the Ursa Major Cluster is not just the nearest to us, it may be the core of the still larger and looser Ursa Major Stream that our solar system is currently passing right through!

MORE DIFFERENCES BETWEEN GLOBULAR AND OPEN CLUSTERS

We've already established that globular clusters contain thousands of times more stars, and are much more densely packed, than open clusters, and also that they are much fewer and farther from us than open clusters. Astronomers, however, have determined some other properties of fundamental importance about these two types of cluster.

If we plot the positions of all known globular clusters, for instance, we discover that the globulars are arranged in a kind of vast spherical halo around the center of the galaxy. Whereas most of the galaxy's stars are either in the enormous central hub or in the flattened outer disk that contains the spiral arms, the globulars can be in any direction from center and at great distances. (See Figure 37.)

The globulars are in some ways like small versions of the central hub of the galaxy: composed almost entirely of ancient, similar stars. In contrast, open clusters are found mostly in the spiral arms of our galaxy (and other galaxies), and like the other stars in those arms they can come in a wide variety of luminosities and spectral types. Most of them are second- or third-generation stars, born from the dust and gas expelled from ancient supernovae and other stars that expelled this material in their death throes.

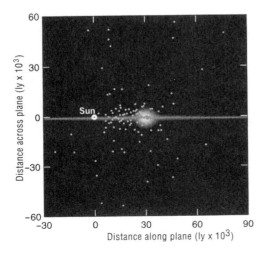

Figure 37.
Galactic halo of globular clusters. (Adapted from diagram in *Astronomy: The Evolving Universe* by M. Zeilik, 7th ed.)

TABLE 9
Selected Star Clusters (Open and Globular)

Open Clusters

Cluster	RA h	m	Dec. °	′	Constellation	Mag.	Size ′	Con.
NGC 869 (h Persei)[a]	2	19	+57	09	Perseus	4.3	30	4
NGC 884 (Chi Persei)	2	22	+57	07	Perseus	4.4	30	3
M34 (NGC 1039)	2	42	+42	47	Perseus	5.5	35	2
Alpha Persei cluster	3	22	+49		Perseus	1.2	185	—
The Pleiades (M45)	3	47	+24	07	Taurus	1.2	110	1
NGC 1502	4	08	+62	20	Camelopardalis	5.7	8	3
The Hyades	4	27	+16		Taurus	0.5	330	1
Collinder 464	5	22	+73		Camelopardalis	4.2	120	—
M38 (NGC 1912)	5	29	+35	50	Auriga	6.4	21	3
M36 (NGC 1960)	5	36	+34	08	Auriga	6.0	12	4
M37 (NGC 2099)	5	52	+32	33	Auriga	5.6	24	4
M35 (NGC 2168)	6	09	+24	20	Gemini	5.3	28	3
NGC 2232	6	27	−04	45	Monoceros	3.9	30	—
NGC 2244	6	32	+04	52	Monoceros	4.8	24	1
NGC 2264	6	41	+09	53	Monoceros	3.9	20	1
M41 (NGC 2287)	6	47	−20	44	Canis Major	4.6	38	3
M50 (NGC 2323)	7	03	−08	20	Monoceros	5.9	16	3
M47 (NGC 2422)	7	37	−14	30	Puppis	4.5	30	2
M46 (NGC 2437)	7	42	−14	49	Puppis	6.1	27	4
M93 (NGC 2447)	7	45	−23	52	Puppis	6.2	22	5
NGC 2451	7	45	−37	58	Puppis	2.8	45	—
NGC 2477	7	52	−38	33	Puppis	5.7	27	5
The Beehive (M44) (NGC 2632)	08	40	+20	00	Cancer	3.1	95	2
M67 (NGC 2682)	08	50	+11	49	Cancer	6.9	30	4
Coma Star Cluster	12	25	+26		Coma Berenices	1.8	275	1
NGC 6231	16	54	−41	48	Scorpius	2.6	15	3
M6 (NGC 6405)	17	40	−32	13	Scorpius	4.2	15	3
IC 4665	17	46	+05	43	Ophiuchus	4.2	41	1
M7 (NGC 6475)	17	54	−34	49	Scorpius	3.3	80	3
M23 (NGC 6494)	17	57	−19	01	Sagittarius	5.5	27	3
M21 (NGC 6531)	18	05	−22	30	Sagittarius	5.9	13	2
NGC 6530	18	05	−24	20	Sagittarius	4.6	15	3
M24[b]	18	17	−18	29	Sagittarius	4.5	90	—
M16[c] (NGC 6611)	18	19	−13	47	Serpens	6.0	35	1

Open Clusters

Cluster	RA h	m	Dec. °	′	Constellation	Mag.	Size ′	Con.
NGC 6633	18	28	+06	34	Ophiuchus	4.6	27	2
M25 (IC 4725)	18	32	−19	15	Sagittarius	4.6	32	2
M11 (NGC 6705	18	51	−06	16	Scutum	5.8	14	5
The Coathanger[d]	19	25	+20	11	Vulpecula	3.6	60	—
M39 (NGC 7092)	21	32	+48	26	Cygnus	4.6	32	3

Globular Clusters

Cluster	RA h	m	Dec. °	′	Constellation	Mag.	Size ′	Con.
47 Tucanae	00	24	−72	05	Tucana	4.5	31	3
NGC 5139 (Omega Centauri)	13	27	−47	29	Centaurus	3.7	36	8
M3 (NGC 5272)	13	42	+28	23	Canes Venatici	6.4	16	6
M5 (NGC 5904)	15	19	+02	05	Serpens	5.8	17	5
M80 (NGC 6093)	16	17	−22	59	Scorpius	7.2	9	2
M4 (NGC 6121)	16	24	−26	32	Scorpius	6.0	26	9
M13 (NGC 6205)	16	42	+36	28	Hercules	5.9	16	5
M92 (NGC 6341)	17	17	+43	08	Hercules	6.5	11	4
M22 (NGC 6656)	18	36	−23	54	Sagittarius	5.1	24	7
M55 (NGC 6809)	19	40	−30	58	Sagittarius	7.0	19	11
M15 (NGC 7078)	21	30	+12	10	Pegasus	6.4	12	4
M2 (NGC 7089)	21	34	−00	49	Aquarius	6.5	13	2

Note: The RA and declination are in 2000.0 coordinates; the magnitude is visual but approximate (sometimes made deceptively bright by the presence of one bright star); the size is usually photographic and thus for visual purposes quite approximate; "Con." is the concentration on a 1 to 5 scale (5 being most concentrated) for open clusters, and on a 1 to 12 scale (12 being most concentrated) for globular clusters.

[a] h Persei and Chi Persei together are known as the Double Cluster.

[b] M24 is the Small Sagittarius Star Cloud.

[c] M16 is here a star cluster but is better known as the nebula (the Star-Queen Nebula or Eagle Nebula) associated with these stars.

[d] Also known as Brocchi's cluster and Collinder 399.

KINDS OF NEBULAE

A nebula is a cloud of interstellar gas and dust that may be giving birth to new stars or being produced by a dying star; it may be lit by nearby starlight, or caused to produce its own light by radiation from nearby stars, or it may be dark.

Diffuse nebulae are ones of irregular shape in which stars may be formed. The most famous examples are the Great Nebula in Orion and the Lagoon Nebula. Diffuse nebulae can be further subdivided into **emission nebulae** and **reflection nebulae**. Emission nebulae actually produce light of their own when their gas is stimulated to fluoresce by the ultraviolet radiation from very hot stars nearby or within. On photographs, a red emission from hydrogen gas is prominent. In the Orion nebula the very hot stars of Theta Orionis also produce an emission of doubly ionized oxygen that the eye is more sensitive to; thus the predominant color of this nebula in telescopes is a shade of green. Reflection nebulae appear bluish, though this color may be only faintly noticed by some observers even in fairly large amateur telescopes. Their light is simply light reflected off the nebula from nearby stars.

Hyades, Pleiades, and Saturn (lower right). (Photographed one Saturn-orbit ago by Steve Albers)

The Merope Nebula and other faint nebulosity that can be glimpsed around some of the Pleiades is a reflection nebula. (The Pleiades are not quite hot enough to cause emissions in these gases.) Some large nebulae have both emission and reflection areas.

Dark nebulae are clouds of gas and dust that neither reflect nor emit light. They may be seen by the absence of stars where they lie, or by their dark silhouette against a more distant bright nebula. An example of the latter is the famous but visually difficult Horsehead Nebula in Orion, a dark notch in front of a glowing strand of nebula seen in medium-size to large amateur telescopes in excellent sky conditions.

Planetary nebulae are so named because many of them appear roughly like the outer planets Uranus and Neptune in telescopes: roundish, small, green or blue, and of rather high surface brightness. Such nebulae are created when material is ejected from a very hot, dying "central star." The bigger, brighter planetary nebulae can take on a variety of shapes in the telescope, as is demonstrated by the names given them, like the Ring Nebula, the Dumbbell Nebula, the Helix Nebula, and the Saturn Nebula. Of these, only the Helix—the biggest of all planetary nebulae—has a low surface brightness. In general, planetary nebulae are of quite high surface brightness: they are often recommended as targets for observers with telescopes in cities. The only problem is that most planetary nebulae are small in apparent size, sometimes just a few arc seconds across. Consequently, you may have to use higher magnification to identify which of the points of light in your telescope are planetary nebulae. But

Large Magellanic Cloud and Tarantula Nebula. (Photographed by Akira Fujii)

remember that the higher your magnification, the narrower your field of view. So you must be sure you are looking very close to the right spot for your planetary nebula or it might not even be in your field. At least the strong blue or green of a planetary nebula is sometimes apparent enough at low magnification to tip you off as to its identity.

SUPERNOVA REMNANTS

Imagine seeing the shimmering wisps of a star explosion that took place tens of thousands of years ago. That is what telescope users behold when they detect, in dark sky conditions,

Small Magellanic
Cloud and
47 Tucanae.
(Photographed by
Akira Fujii)

TABLE 10

Selected Planetary Nebulae

Nebula	RA h	m	Dec. °	'	Constellation	Mag. C. Star	Mag. Neb.	Size
M76 (NGC 650-1) (Little Dumbbell Nebula)	01	42	+51	34	Perseus	16–17	12?	120 × 60
NGC 2392 (Eskimo Nebula)	07	29	+20	55	Gemini	10	8.3	40
NGC 3242 (The Ghost of Jupiter)	10	25	−18	38	Hydra	11.5	8	40
M97 (NGC 3587) (Owl Nebula)	11	15	+55	01	Ursa Major	14	11	150
M57 (NGC 6720) (Ring Nebula)	18	54	+33	02	Lyra	14–17 (var.?)	9	70 × 150
NGC 6826 (the Blinking Planetary)	19	45	+50	31	Cygnus	11	9	25
M27 (NGC 6853) (Dumbbell Nebula)	20	00	+22	43	Vulpecula	13.5	8.1	480 × 240
NGC 7009 (Saturn Nebula)	21	04	−11	22	Aquarius	12	8.4	26
NGC 7293 (Helix Nebula)	22	30	−20	48	Aquarius	13	6.5	900 × 720

The RA and declination are in 2000.0 coordinates; the magnitudes may be approximate; the size (given in arc-seconds) is usually photographic and thus for visual purposes quite approximate.

the various sections of the Veil Nebula in Cygnus. The Crab Nebula, in Taurus, caused by a supernova that flared brighter than Venus in the year 1054, is an even more famous example of a **supernova remnant** (**SNR**). (By the way, it's fascinating to consider that even the Crab Nebula, seen relatively recently in human history, actually occurred several thousand years before 1054—it took the light from the event that long to reach us.)

Scientists do not fully understand why some supernovae fail to leave prominent remnants for hundreds or thousands of years. But the Crab and the Veil are fairly conspicuous and are among the most intriguing objects in all the heavens.

The beginner, perhaps having seen long-exposure photographs of these nebulae, is often disappointed by a first visual inspection, especially if it is with a small telescope or in mediocre sky conditions. But keep trying! The proper use of averted vision can make a

TABLE 11

Selected Diffuse Nebulae and Supernova Remnants

| Nebula | RA | | Dec. | | Constellation | Mag. | Size |
	h	m	°	'			'
M1 (NGC 1952) (Crab Nebula)[a]	5	34	+22	01	Taurus	8.2	6 × 4
M42 (NGC 1976) (Great Orion Nebula)	5	35	−05	35	Orion	2.9	66 × 60
NGC 1977	5	35	−04	52	Orion	4.6	20 × 10
M43 (NGC 1982)	5	36	−05	16	Orion	6.9	20 × 15
NGC 2237 (Rosette Nebula)	6	32	+05	03	Monoceros	—	80 × 60
M20 (NGC 6514) (Trifid Nebula)	18	03	−23	02	Sagittarius	8.5	29 × 27
M8 (NGC 6523) (Lagoon Nebula)	18	04	−24	23	Sagittarius	5	90 × 40
M16 (NGC 6611) (Star-Queen Nebula [Eagle])	18	19	−13	47	Serpens	6.0[b]	35
M17 (NGC 6618) (Omega Nebula)	18	21	−16	11	Sagittarius	7	46 × 37
Veil Nebula[a]							
NGC 6960	20	46	+30	43	Cygnus	—	70 × 6
NGC 6992	20	56	+31	43	Cygnus	—	60 × 8
NGC 7000 (North American Nebula)	20	59	+44	20	Cygnus	—	120 × 100

Note: The RA and declination are in 2000.0 coordinates; the magnitude is approximate; the size is usually photographic and thus for visual purposes quite approximate.

[a] Supernova remnant.

[b] Magnitude of the associated cluster.

tremendous difference in how much can be glimpsed. And an eight-inch or ten-inch tele-scope in dark skies can really begin to show the exquisitely lovely thin filaments in both of these remnants, as well as the scalloped edges of the Crab Nebula.

SUMMARY

In addition to densely packed globular star clusters, there are much looser galactic or open star clusters. Even looser is the arrangement of stars known as a moving group, while young stars rapidly moving out of an area of relatively recent star formation remain a kind of star association. Globular clusters are distributed in a vast spherical halo located around the center of the Milky Way galaxy, whereas open clusters are mostly spread through the equa-torial disk—especially the spiral arms—of this and other galaxies. Globular clusters contain mostly similar, first-generation stars, whereas the stars in spiral arms are mostly second- or third-generation and of greater variety.

Diffuse nebulae are irregularly shaped clouds of dust and gas from which stars may form. Those that shine only by reflecting starlight are called reflection nebulae. Those that are stimulated to glow on their own by ultraviolet light of very hot stars are called emission nebulae. Dark nebulae are clouds of dust and gas not illuminated by nearby stars but rather seen by the scarcity of stars in front of them or in silhouette against a bright nebula. Planetary nebulae are the ejected gas from dying central stars, stimulated to glow by the ultraviolet light of the central star. Supernova remnants are the various kinds of glowing clouds from past supernova blasts.

◆ NIGHT 37 ◆

GALAXIES

◆ ◆ ◆

Time: Any night.

BEYOND OUR MILKY WAY float billions of other galaxies. Most are not as big and bright as ours. But each is an "island universe": a vast congregation of usually at least billions of stars.

Galaxies are so distant that the light we see from them tonight left them millions of years ago, or even billions of years ago—a significant fraction of the age of the universe. Not surprisingly, then, few galaxies are visible to the naked eye, and even telescopes can show only glimpses of structure in most of them.

But what a thrill it is to see such things at all! According to the song, "On a clear day you can see forever." However, on a clear night in autumn your naked eye really can allow you to see almost three million years into the past (the Great Andromeda Galaxy is almost three million light-years away), and on a clear night in spring (in other seasons, too) a small telescope can permit you to see fifty, sixty, or seventy million years in the past: the images of galaxies as they appeared when the dinosaurs were still alive on Earth. Larger amateur telescopes can show both the specks of galaxies and the mysterious objects called "quasars" as they looked up to several billion years ago.

The Galaxies of Spring and Autumn

The gas and dust in the equatorial plane of our Milky Way galaxy block our view of most other galaxies that lie fairly close to the Milky Way band in our sky. That band crosses the southern and overhead sky in summer and winter, so galaxies appear extremely scarce among the traditional constellations of those seasons.

In contrast, when we look toward many of the traditional constellations of spring and autumn, we are staring straight "up" or "down" from the equatorial disk of the Milky Way, looking through the least distance of dust and gas possible. Consequently, we are able to detect vast numbers of galaxies, and some of them are bright enough to see fairly easily in small telescopes.

Observe on various nights as many of the galaxies of spring and autumn listed in Table 12 as you can (refer to Figure 39 for one detailed map).

Sometimes it is good to know what to expect before you actually look, though looking first provides a fresher and less biased view. A case in point is galaxies, which in small telescopes may show mere traces of structure—so elusive that you really have to know what you're looking for to detect it.

The most important categorization of galaxies is according to structure. There are three basic types: spiral galaxies, elliptical galaxies, and irregular galaxies. Irregular galaxies have irregular shapes. Some were no doubt once spirals or ellipticals that were disrupted by the pull of bigger nearby galaxies or by actual collisions with other galaxies. Elliptical galaxies range from round to elongated ovals and seem to resemble—and have the old stars typical of—the central hub of spiral galaxies. Some elliptical galaxies are far huger than any spiral galaxy's hub, however. Spiral galaxies are the most intricate in form, with their central hubs and surrounding pinwheel-like disks of spiral arms. Those arms are often ablaze with

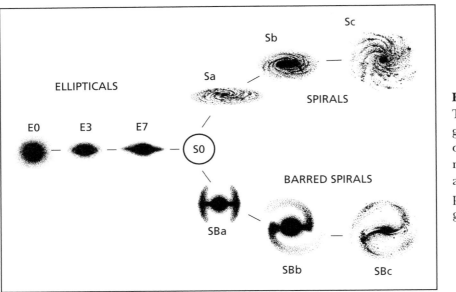

Figure 38. Types of galaxies (this diagram doesn't mean to imply an evolutionary progression of galaxy forms).

bright, hot new stars and their surrounding luminous nebulae, but they are also riddled with dark gas and dust. Our Milky Way is a spiral galaxy, as is its much bigger neighbor, the Great Andromeda Galaxy, M31.

Each of these major classes can be divided into further subclasses or types. Spiral galaxies are divided into types according to how tightly wrapped their spiral is and whether or not there is a bright bar of material across it. In addition, we see the spirals tilted at various angles to us. Some we see almost face-on; they look like pinwheels. Others appear virtually edge-on and resemble sombreros or needles, usually brimmed or split with a lane of dark gas and dust.

ANDROMEDA'S GALAXIES AND OTHERS

Autumn offers the brightest external galaxy visible from the northern middle latitudes of Earth. That is the mighty M31 in Andromeda, which we discussed as a naked-eye object earlier in this book. Actually, the naked eye or binoculars in very dark skies are better than telescopes at detecting the low surface brightness at the ends of this galaxy. Your unaided eyes might see the elongated smudge of M31 up to 4° or even 5° long under excellent conditions. But telescopes show much better the intense central hub of the galaxy and, if your telescope is big enough and "seeing" good enough, an even more intense nucleus. The spiral of M31 is gently tipped towards us. A six-inch or eight-inch telescope can easily show the

TABLE 12

Selected Galaxies

Galaxy	RA h	m	Dec. °	′	Constellation	Mag.	Size ′
NGC 55	00	15	−39	11	Sculptor	8	32 × 6
M110 (NGC 205)	00	15	+41	41	Andromeda	8.0	17
M32	00	43	+40	52	Andromeda	8.2	8 × 6
M31[a]	00	43	+41	16	Andromeda	3.5	160 × 40
NGC 253	00	48	−25	17	Sculptor	7.1	22 × 6
M33[a]	01	34	+30	39	Triangulum	6.3	60 × 35
M81	09	56	+69	04	Ursa Major	7.0	26 × 14
M82	09	56	+69	41	Ursa Major	8.4	11 × 5
M49	12	30	+08	00	Virgo	8.4	9 × 7
M87	12	31	+12	24	Virgo	8.6	7
M104[a]	12	40	−11	49	Virgo	8.3	9 × 4
M94	12	51	+41	07	Canes Venatici	8.2	11 × 9
M64[a]	12	57	+21	41	Coma Berenices	8.5	9 × 5
NGC 5128[a]	13	26	−43	01	Centaurus	7	18 × 14
M51[a]	13	30	+47	12	Canes Venatici	8.4	11 × 8
M83	13	37	−29	52	Hydra	7.6	11 × 10
M101	14	03	+54	21	Ursa Major	7.7	27 × 26

Note: The selected galaxies listed are selected from those visible from 40° N latitude.
[a] M31 is Great Andromeda Galaxy; M33 is Pinwheel Galaxy; M104 is Sombrero Galaxy; M64 is Black-Eye Galaxy; NGC 5128 is Centaurus A; M51 is Whirlpool Galaxy.

brightest star cloud and traces of dark lanes running parallel to the direction in which the galaxy is elongated.

Such a telescope (and smaller) will also show M31's two most important companion galaxies, M32 and M110. (M110 is often referred to as NGC [New General Catalogue] 205, for some critics refuse to accept it as being legitimately a member of Charles Messier's catalogue of deep-sky objects.) These two are both elliptical galaxies, similar in total brightness, but M32 is more concentrated, rounder, and appears closer to M31.

During the months when Andromeda's galaxies are highest in the evening sky, observers at southerly latitudes have their best evening view of the Milky Way's two most important companion galaxies, the Large Magellanic Cloud and the Small Magellanic Cloud. These irregular galaxies look like detached pieces of Milky Way. They are both fewer than 200,000 light-years away, compared to M31's distance of almost three million light-years.

Another member of our "local group" of galaxies is in Andromeda's neighbor constellation, Triangulum. The spiral galaxy M33 is barely visible to a keen naked eye on very good nights at the darkest observing sites. Try for it with quite low magnification, since spreading out the image of this low-surface-brightness galaxy makes it much harder to see.

There are many galaxies of interest in Cetus, Pisces, and Eridanus, but brighter are several in the constellation Sculptor. NGC 253 is a pretty spiral when the night is clear and when it is at its height in the south, especially when viewed through an eight-inch or larger telescope.

Spring and Its "Realm of the Galaxies"

There are clusters of galaxies in space. The most famous is the great Virgo Cluster of spring. This helps make a region centered in northwestern Virgo deserve the title "Realm of the Galaxies." Figure 39 shows a close-up map you can use to seek out some of the dozens of galaxies a medium-size telescope can show here in dark skies. If you have such a telescope and conditions, here is where you can get the greatest number of galaxies in one night, one hour, or one field of view.

M49 and M87 are big and, of course, relatively bright members of the Virgo Cluster, which may be centered about forty or fifty million light-years away. (The light you see from them tonight left them almost as far back as the last days of the dinosaurs.) But there are other, even more prominent galaxies in the spring sky that may be easier to find.

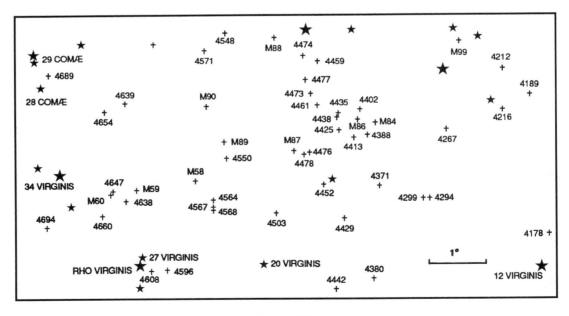

Figure 39.
The region of sky containing the central part of the Virgo galaxy cluster
(four-digit numbers have had the "NGC" before them removed to reduce clutter).

Far north in Ursa Major are the bright odd couple M81 and M82. Separated by little more than half a degree, they offer a spectacular contrast: M81 is a beautiful, regularly formed spiral, M82 a bizarre cigar-shaped object. Near the opposite corner of Ursa Major, just under the end of the Big Dipper's handle in Canes Venatici, is M51, the Whirlpool Galaxy. M51 offers observers the easiest chance to see full-fledged spiral structure. A six-inch telescope will sometimes suffice, and larger amateur telescopes show that it seems to be attached by a large spiral arm to an easily visible patch of light—which turns out to be another, peculiar galaxy (NGC 5195).

Our table lists a few other bright galaxies of spring. The one perhaps most worthy of note is M104, the Sombrero Galaxy, the brightest-appearing of the galaxies that are presented to us nearly edge-on. A medium-size telescope starts to show well the dark lane that decorates the imagined Mexican hat. M104 is technically in Virgo but is close to the main pattern of Corvus the Crow.

SUMMARY

Billions of galaxies exist beyond our Milky Way. Almost all of them are many millions or even billions of light-years away. Three major kinds of galaxy are spiral, elliptical, and

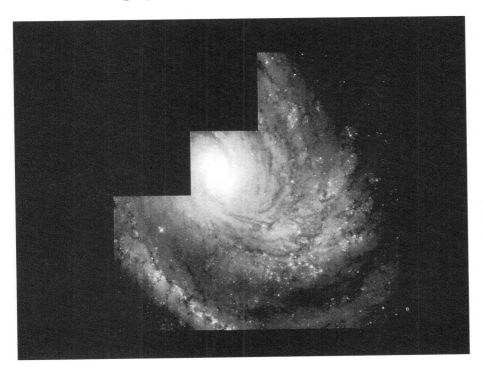

Part of the M100 Galaxy. (Photograph by Hubble Space Telescope)

irregular. Each kind is further classified into subtypes. Spiral galaxies are classified according to how tightly their spiral arms are wound and whether or not they are crossed with a luminous bar. The constellations of spring and autumn offer the most galaxies visible because these regions are not so much dimmed by the dust and gas in the central plane of the Milky Way. Great autumn galaxies include M31 and its companions in Andromeda and M33 in Triangulum. Spring presents us with the "Realm of the Galaxies" caused by the presence of the Virgo Cluster and with other fine galaxies like M81 and M82, M51, and M104.

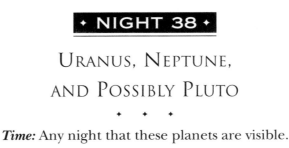

✦ NIGHT 38 ✦

URANUS, NEPTUNE,
AND POSSIBLY PLUTO

✦   ✦   ✦

Time: Any night that these planets are visible.

VENUS IS SO BRIGHT that it is difficult for people to miss when it is in the evening sky. Jupiter is the most common bright light through large parts of the night. Mars occasionally flares up to spectacular brightness and ruddy prominence, and the rest of the time it can be found, like Saturn, without much difficulty as one of the one or two dozen brightest points of light in the sky. Mercury is more elusive, prominent only on clear nights during a few weeks of the year, but it's certainly detectable to the naked eye if you make a point of learning when and where to look.

These five planets, however, are the ones which have been known throughout history (and surely before). The final, farthest, faintest planets—Uranus, Neptune, and Pluto—were not identified until telescopes—and, in Pluto's case, photography—could reveal them in the eighteenth century (Uranus), nineteenth century (Neptune), and twentieth century (Pluto). Needless to say, this means that the backyard astronomer needs some special knowledge and, in most cases, optical assistance to complete his or her list of observed planets.

WHAT YOU NEED TO SEE THEM

Uranus lies about twice as far out from the Sun and Earth as Saturn does. Yet Neptune is about a billion miles even farther out, and Uranus is only halfway as far out as the average distance of Pluto. Fortunately, Uranus and Neptune are gas giants (four times wider than Earth but little more than a third the diameter of enormous Jupiter), so they reflect enough light to see Uranus dimly with the naked eye under excellent sky conditions and Neptune with binoculars. Pluto is so tiny—less than half the size of Mercury, much smaller than Earth's Moon—that it is usually too dim to see in anything less than a fairly large amateur telescope. Not even the world's largest ground-based telescope really shows it as larger than a point of light.

The globes of bigger, closer Uranus and Neptune are small enough at low magnification that they are quite hard to find without a detailed finder chart. In other words, you must use a low-magnification eyepiece to have a wide view and check your chart to see which one of the points of light is Uranus or Neptune. Under good conditions you may notice that their color is a peculiar bluish green; this is far more noticeable in the brighter Uranus. But don't count on finding the planet without a chart.

Such finder charts appear every year in magazines like *Sky & Telescope* and *Astronomy* (usually in the April or May issue of the former, for nowadays April or May is the first month of the year these planets are well enough placed for convenient observation). Finder charts for Uranus, Neptune, and Pluto also appear in some annual publications, like Guy Ottewell's *Astronomical Calendar.*

Understand that identifying Pluto is several orders of difficulty harder than identifying Uranus and Neptune. Most experts recommend at least a ten-inch telescope and fairly dark skies. This is true even though Pluto is now near the most inward point of its almost 250-year orbit. In the period from 1979 to 1999, Pluto has actually been slightly closer to the Sun than Neptune. Even so, the brightest it has gotten is about magnitude 13.7. Most of the finder charts for it barely reach that limiting magnitude, so for that and other reasons the proper way to identify Pluto is to observe it on two different nights, ascertaining that the point of light you are seeing really has changed position. The basic instructions for observing Pluto are: *Use a detailed finder chart and a telescope of preferably ten inches or more aperture, and try to identify the speck of light that is Pluto on two separate nights.*

OBSERVING URANUS AND NEPTUNE

Not many amateur astronomers have seen Pluto, and very few indeed have seen it multiple times. But there is no reason why anybody with charts and a small telescope can't easily and frequently observe sixth magnitude Uranus and eighth magnitude Neptune if they want to.

Much of the pleasure of seeing these worlds lies in merely knowing what you are seeing. Besides their interesting color, all that you are likely to see without a large telescope, great conditions, and much observing experience is simply the globes of these worlds.

Features in the clouds of Uranus and Neptune are sometimes suspected by amateur observers, but anything prominent enough to defend or sketch is a real rarity. At least the globe of Uranus is easy to detect. On a night of reasonably good "seeing," the nearly four arc-second-wide globe of Uranus starts becoming detectable at about 50×. If conditions and telescope permit, the sight of a pea-sized Uranus at higher powers is delightful.

With Neptune, the more common challenge is just to make certain you are viewing a globe rather than a fat, shaky starlike object. At its distance of more than two and a half billion miles from Earth, Neptune's disk subtends little more than two seconds of arc.

The basic instructions for observing Uranus and Neptune are: *Use a finder chart and binoculars or telescope to locate Uranus and Neptune at low power. Then switch to higher powers to try to get the best view possible of the globes of the colorful planets.*

SUMMARY

Special preparation is needed to observe Uranus, Neptune, and Pluto. Uranus can be glimpsed with the naked eye under very good conditions, but a finder chart and binoculars are normally required to locate Uranus and always to locate the dimmer Neptune. Pluto is so dim that it can be located only with special finder charts and a telescope, preferably one with ten inches or more of aperture. Much smaller amateur telescopes can show the color of Uranus and Neptune and, on nights of good "seeing," their tiny apparent disks.

✦ NIGHT 39 ✦

COMETS AND ASTEROIDS

✦ ✦ ✦

Time: Any time that one or more comets or asteroids are visible.

ANYONE WHO SAW Comet Hale-Bopp in its climax months of 1997—as did perhaps as many as 80 percent of all Americans—will not need to be told that great comets are one of the sky's most amazing spectacles. This is even more true for people who saw Hale-Bopp in a good telescope or saw either it or 1996's electrifying blue Comet Hyakutake in dark rural skies. Of the more or less predictable sights in the heavens, only a total eclipse of the Sun is more staggeringly awesome—but it lasts no more than a few minutes, while a great comet can give us a rich array of sights for weeks or even months. And a total eclipse of the Sun is visible over only a small percent of Earth's surface, while half or sometimes even all the world gets to enjoy a major comet.

So why haven't we dealt with comets earlier in this quest to know the night sky? Because, alas, bright comets do not appear very often, and the more typical faint comets are in many ways not an easy target for beginners, even with telescopes. Those other so-called minor members of the solar system, the asteroids, are almost all too faint to be seen without a telescope, and, unlike comets, they look much like stars. To identify them you need finder charts that are usually more detailed than those for Neptune.

Yet even a glimpse of a dim comet or asteroid is a most desirable goal. After all, asteroids and, even more so, comets, offer immense variety in their great numbers. And solving their mysteries is more likely than anything else to teach us about the very origins of the solar system.

THE DEEP MYSTERIES OF COMETS

Before discussing how to observe comets and what they show us, we should learn some of the more remarkable facts—and unknowns—about their nature.

For almost all of history, scientists were not even close to finding out what comets were or where they came from—not only where they came from in the formation of the solar system but where any particular comets came from before appearing suddenly in the sky.

After all, even in ancient times, skywatchers studied the motions and appearances of the Sun, Moon, planets, and stars well enough to recognize regularities. They did a pretty good job of predicting where these objects would go, even if they didn't know what the objects were and which body—Sun or Earth—they went around. For instance, the Sun, Moon, and planets moved but their motions were confined to that one band of constellations, the zodiac. In contrast, comets could appear anywhere in the heavens, travel anywhere, and disappear anywhere. Planets could dim and brighten considerably, but they did so according to clearly discernible patterns. Comets could brighten a hundredfold, fade, then brighten again. And they didn't appear as mere points of light like planets and stars do to the naked eye. Bright comets usually had a tail trailing from a larger than point-size patch of fuzzy light which was the head of the comet. They were the outcasts, the oddballs, the renegades of the solar system, and their eerie appearance compelled the superstitious to think that this disruption of the order in the heavens would be followed by disruption on Earth: plagues, wars, the deaths of rulers.

It wasn't until 1705, when Edmond Halley published his most famous paper, that humankind started learning that some of the comets they saw were the same ones returning and that some of these orbits were highly elongated. As more comet orbits were calculated, it was found that some had their most distant point tremendously far beyond Pluto; others had their near point so close to the Sun that they literally skimmed the blinding solar surface.

But scientists still didn't have an inkling of the ultimate source of comets. And they were not at all sure what solid body or bodies might exist in the midst of the vast clouds of glowing dust and gas that billowed out as a comet got closer to the Sun. It wasn't until 1950 that two big new theories laid the groundwork for the answers.

One theory was astronomer Fred Whipple's. He predicted that there is a solid object at a comet's center, known as its **nucleus**, composed of ice (mostly water ice) shot through with dust and rock particles. Whipple's **dirty snowball model** was actually more complicated and became much more refined over the following decades. It was essentially confirmed by photographs of Comet Halley's nucleus taken by the two Vega and one Giotto spacecraft that passed near the nucleus in March 1986.

The other theory was that of the **Oort Cloud**.

OORT, KUIPER, AND LONG- AND SHORT-PERIOD COMETS

In 1950, astronomer Jan Oort proposed that there exists a vast cloud of at least billions of pristine comet nuclei at tremendous distance beyond the planets, a sizable fraction of the way to the nearest star system. Some of these comet nuclei would be found more than one light-year from the Sun. When stars came unusually close to the solar system, they would disrupt countless thousands of comets (maybe millions), the gravitational tugs speeding up some comets and slowing down others. This would result in many comet nuclei being ejected away from the solar system but also many sent dropping in toward the planets, even toward the inner solar system, where the comet ice would vaporize as it neared the Sun and produce the cloud of head and tail that make a comet more readily visible from Earth. This first-time visitor to the inner solar system would curve around the Sun and hurtle off to an aphelion (far point from the Sun) back out in the Oort Cloud. But the comet would now be on a tremendously elongated orbit that would bring it back to the neighborhood of Earth and Sun. At one of its visits, it would pass near enough to a planet—probably powerful Jupiter—for the planet's gravity to alter its orbit, possibly into a much smaller one, merely thousands of years long.

The Oort Cloud would explain how there still seems to be, four and a half billion years into the history of the solar system, an influx of new comets into the inner solar system. But in the 1990s there has come dramatic support for a theory raised by Gerard Kuiper in the 1940s, that not all of our comets are derived from the Oort Cloud. One major kind of comet, we now think, does come from the Oort Cloud, but another major kind comes from a more flattened disk-shaped region of comets much closer to the orbits of the outermost planets.

Many decades of comet-orbit calculations after Edmond Halley's showed that comet orbits fell into two groups: **short-period** (or **periodic**) **comets**, the ones with orbital periods of less than two hundred years, and **long-period comets**, the ones with orbital periods of more than two hundred years. Some small, rather dim comets return every six, five, even three years, completing their orbits (which go out to near Jupiter's orbit) in that time. One of the short-period comets with the longest period is the first ever predicted to return: Comet Halley. Halley's is the brightest comet we get to see well about once in a long human lifespan, once in about seventy-five years. Much brighter than Halley's, in true and sometimes apparent brightness, are some of the long-period comets, like Hale-Bopp. Most of the long-period comets have periods very much longer than two-hundred years—thousands or even millions of years long.

Until recently, it was believed that short-period comets were just long-period comets from the Oort Cloud whose periods had been shrunk drastically by encounters with the gravitationally powerful planets. But explorations of the orbital mechanics of how this could happen began to make it seem unlikely. Then, in the 1990s, astronomers began to discover dozens of fairly small bodies just beyond the orbit of Neptune. There now seems little doubt that these are the first detected members of the **Kuiper Belt**, a flattened belt of multitudes of comets that is the actual supplier of short-period comets.

The comet nuclei in the Oort Cloud and also, presumably, the Kuiper Belt may be virtually unspoiled samples of the material that formed the larger bodies of the solar system more than four billion years ago. So anything we can learn about these two vast comet repositories and about comets themselves could be vital to understanding the origins of our own planet and solar system. There are also theories that Earth's collisions with comets billions of years ago helped establish life on our planet, by bringing water, organic molecules, or both to our world. And there is strong evidence that Earth being hit by comets and/or asteroids has, at intervals of many millions of years, caused tremendous mass extinctions—the most famous being that of the dinosaurs sixty-five million years ago.

OBSERVE A COMET

When you observe a comet, then, you are looking at one of the objects that have been possibly Earth life's greatest creative agents and almost certainly its greatest destroyers. You are looking at the secret clues to the very origin of our solar system and us.

But even if you don't ponder these thoughts, the visual inspection of comets, even the dim ones, is fascinating. Even a dim comet is a strangely changing patch of light, pursuing

Comet West.
(Photographed by
Akira Fujii in 1976)

a distinctive path across the heavens over the course of weeks and months. It's like having a nebula that changes its appearance every night or week—might you be the person in all the world who is getting the best look at it tonight?

You have gotten details about a new comet (see Sources of Information in the back of this book) and know where to look for it tonight. Scanning with your telescope, you suddenly come upon it: a sizable hazy patch of light where none would otherwise be. Try to estimate how concentrated it is toward its center. No concentration—completely diffuse—is 0 on the DC (degree of condensation) scale, while a comet whose head appears starlike gets a 9 on the scale. Most comets fall in between these extremes, and what you are looking at is merely the **coma***, the cloud of gas and dust of the comet's* **head***. The head is composed of both the large coma and its comparatively tiny source, buried deeply within it, the nucleus. Not all comets get active or close enough to Earth and Sun to display the other major region of a full-fledged comet. But in this comet you do glimpse, very dimly, a streamer of glow: the comet's* **tail***. Bright comets often display both a straight, narrow* **gas tail** *(shining by fluorescence caused by atomic particles of the "solar wind" hitting the comet's gases) and a curved, broad* **dust tail** *(shining by reflecting sunlight).*

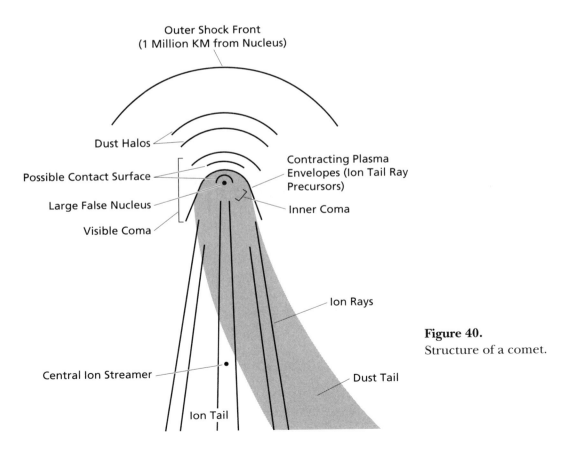

Figure 40.
Structure of a comet.

The comet's ionized gas is driven straight away from the Sun by the Sun's endless outflow of atomic particles, the solar wind. But the dust tail is composed of particles that are pushed away from the Sun more slowly—by the pressure of sunlight itself. Moving more slowly, the dust particles have longer to lag behind the comet's motion and form a curved tail. Which kind of tail are you now seeing? Sometimes we see a comet tail from a side view, with gas and dust tails overlapping each other, in which case the two types of tail are difficult to differentiate.

Very bright comets can show all kinds of wondrous features: a parabolic outline to the coma, color (blue gas and yellowish or reddish dust), "ion ray" streamers in the gas tail, and more. The next time we get a comet even half as brilliant as Hale-Bopp, you'll see them and sketch them!

ESTIMATING COMET BRIGHTNESS

There is one thing you can do with any comet if you know the brightness of comparison stars near it (a star atlas and catalogue gives this information): estimate the comet's own brightness. The problem is that you are trying to compare the brightness of a point of light (a star) with a more extended diffuse one (a comet's coma). One solution is the Bobrovnikoff Method, or Out-Out Method. The observer moves both the comet and stars further and further out of focus until the image of the star looks almost as large as the comet's—then the comparison in brightness is made. This is the easiest way to estimate comet brightness and is reasonably reliable for most comets.

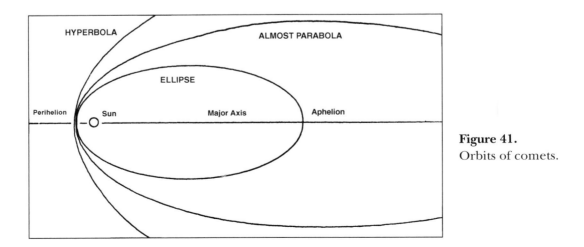

Figure 41.
Orbits of comets.

DON'T FORGET THE ASTEROIDS

In my fervor for comets, I have neglected discussing the asteroids much. I've even made them out to be rather dim and hard to identify for the beginner. But they are truly worth some extra effort. There are thousands of these worlds that are bigger than most comet nuclei, and there is hope that our spacecraft will explore a number of them thoroughly in the next few decades. Because far more of them come closer to Earth, and more frequently, than comets, there are unusual opportunities for amateur astronomers to see them brighten incredibly rapidly, move across whole constellations in hours, and—if they are small, irregularly shaped asteroids—give brightness fluctuations that reveal their orbital period. There may even be opportunities for missions with human crews to explore a few of the closest-passing asteroids in the decades ahead, though it will probably be a while longer before anyone starts mining them! Somebody certainly needs to identify and track those asteroids that already come close to Earth and might—not likely, but just possibly—pose a threat in the next few centuries (or the next year!).

Asteroids more often and more clearly perform visible occultations of stars than comet nuclei do. These occultations also provide opportunities to prove that certain asteroids have their own moons. The Galileo spacecraft photographed such a moon with the asteroid Ida a few years ago.

It's too bad that the wonderful publication *Tonight's Asteroids* no longer exists. But with the help of finder charts and specific information in astronomy magazines and some annual publications (see Sources of Information), you can find great enjoyment in tracking down a wonderful assortment of these varied worldlets as the years go by. The asteroid Vesta sometimes gets bright enough to glimpse with the naked eye in dark skies. The asteroid Ceres is the biggest, 568 miles across. The other members of the Big Four—actually not the biggest four asteroids but the first discovered, at the start of the 19th century—are Pallas and Juno. And there are a number of other asteroids that occasionally get bright enough to see easily in binoculars.

SUMMARY

Comets and asteroids are survivors from among the countless small bodies in the earliest days of the solar system. Comets have been implicated in the formation of life on Earth and in its massive destruction; yet asteroids collide with the Earth more often. After Edmond Halley proved that at least some comets return, many more comet orbits were calculated, and it was found that some comets are short-period or periodic, having orbital periods of

Comet Hale-Bopp displays a broad dust tail to the right and a narrower gas tail to the upper right in this March 1997 photograph by Ray Maher.

less than two hundred years (as little as three years), while others were long-period, having periods of more than two hundred years, and often thousands or even a million years. Comets are nuclei of frozen water ice and other substances, laced with dust and rock, whose surfaces start to vaporize as they approach the Sun on the near ends of their typically very elongated orbits. Most comets are dim, but a telescope shows the coma of gas and dust as a hazy patch of light (the coma and the nucleus within it form the head of the comet) and sometimes a fan or streamer of light pointing more or less away from the Sun—a tail. There are actually two common tails: a straight narrow gas tail and a curved, broad dust tail. Many more structures and color and changes are visible in bright comets, which are among the heavens' most spectacular sights. With any comet, bright or dim, an observer can estimate the degree of condensation toward the center and the brightness. The latter is most easily estimated by moving out of focus the images of stars until they are almost as big as the also out-of-focus comet's and then comparing their brightness.

There are many thousands of asteroids, rocky bodies much smaller than planets. They appear starlike and must be identified with a good finder chart or by watching for motion, usually from night to night. Far more of them come close to Earth than do comets. Asteroids sometimes occult stars; sometimes they possess moons. The first four asteroids discovered were Ceres (the biggest), Pallas, Juno, and Vesta (the brightest).

✦ NIGHT 40 ✦

A FIRST LOOK
AT THE SKY—AGAIN!

✦ ✦ ✦

Time: The first night after you've finished
the previous nights of this book.

YES, IT SOUNDS LIKE a contradiction to talk about taking a first look at the sky *again*. But I mean that the view of the sky you can take now that you have learned what the previous thirty-nine Nights had to offer is not drained of wonder because you understand the sky so much better. Quite the contrary is true.

Of course, there is no way to repeat exactly the experience of the first night you saw a total lunar eclipse, or were able to identify Leo, or watched a meteor shower, or saw the rings of Saturn through a telescope. But these events, patterns, displays, and objects are so rich that you can never exhaust all their possibilities—and you would never have known of their possibilities, perhaps not even become aware of their existence, without your many nights of observation and study.

The heavenly sights still offer many mysteries to intrigue you—in fact, more mysteries, because what you have learned will help you generate new questions.

What do you see when you look at the sky after dark now? Stars in their patterns and their individual varieties of doubleness and variability, reaching the same positions in the sky four minutes earlier each night because the Sun appears to be creeping along the zodiac—all because the Earth is actually pursuing its course around the Sun. Planets continuing their slow eastward trek, while being swept more swiftly west by the effect of Earth's rotation, like all else except the nearby airplane lights, satellites, light pollution glows and meteors. Perhaps a few of the planets appear to be halting and heading westward with respect to the stars—a clear sign to you that such a planet is being overtaken by or is itself overtaking the Earth. Maybe you see the Moon this evening, making eastward progress in its orbit by one apparent diameter of itself per hour in relation to the background stars. Or maybe you don't see it in the sky right now, but if you know either what its phase is, or when it will rise or where it will be, you immediately can figure out the rest.

Perhaps the Milky Way band is visible in your sky. The Big Dipper must be, taking its circumpolar path around the North Star while the stars of your current season have risen in the east and will set in the west, their vaster circles around the celestial poles cut off by your horizon. You know where among the numerous constellations the teeming star clusters and eerily glowing nebulae lie. And as a piece of space rock, a meteor, which once came out of a comet you may see later this year, cuts a fiery path in the atmosphere fifty miles overhead, it passes by a galaxy of billions of stars that will enter your eye through your telescope after traveling fifty million years to reach you.

Your forty nights to knowing the sky have been only the first of many more that you will enjoy in an ever-richening flight of knowledge and beauty.

◆ APPENDIX 1 ◆

SOME BASIC DATA ON THE BRIGHTEST STARS

Star	Constellation	App. Mag.	Abs. Mag.	Distance light-years
Sirius	Canis Major	−1.46	1.4	8.6
Canopus	Carina	−0.72	−2.5	74
Alpha Centauri	Centaurus	−0.27	4.1	4.3
Arcturus	Bootes	−0.04	0.2	34
Vega	Lyra	0.03	0.6	25
Capella	Auriga	0.08	0.4	41
Rigel	Orion	0.12	−7.1	910
Procyon	Canis Minor	0.38	2.6	11.4
Achernar	Eridanus	0.46	−1.3	69
Betelgeuse	Orion	0.50v	−5.6	540
Beta Centauri	Centaurus	0.61v	−4.4	320
Alpha Crucis	Crux	0.76	−4.6	510
Altair	Aquila	0.77	2.3	16
Aldebaran	Taurus	0.85v	−0.3	60
Antares	Scorpius	0.96v	−4.7	440
Spica	Virgo	0.98v	−3.2	220
Pollux	Gemini	1.14	0.7	40
Fomalhaut	Piscis Austrinus	1.16	2.0	22
Beta Crucis	Crux	1.25v	−4.7	460
Deneb	Cygnus	1.25	−7.2	1,500
Regulus	Leo	1.35	−0.3	69
Adhara	Canis Major	1.50	−4.8	570
Castor	Gemini	1.57	0.5	49

Note: The letter v indicates that the star is variable (of these stars, only Betelgeuse and Antares vary markedly). The magnitudes for double stars are the combined brightnesses.

✦ APPENDIX 2 ✦

THE CONSTELLATIONS

Name	Genitive	Abbreviation	Order of Size
Andromeda	Andromedae	And	19
Antlia	Antliae	Ant	62
Apus	Apodis	Aps	67
Aquarius	Aquarii	Aqr	10
Aquila	Aquilae	Aql	22
Ara	Arae	Ara	63
Aries	Arietis	Ari	39
Auriga	Aurigae	Aur	21
Boötes	Boötis	Boo	13
Caelum	Caeli	Cae	81
Camelopardalis	Camelopardalis	Cam	18
Cancer	Cancri	Cnc	31
Canes Venatici	Canum Venaticorum	CVn	38
Canis Major	Canis Majoris	CMa	43
Canis Minor	Canis Minoris	CMi	71
Capricornus	Capricorni	Cap	40
Carina	Carinae	Car	34
Cassiopeia	Cassiopeiae	Cas	25
Centaurus	Centauri	Cen	9
Cepheus	Cephei	Cep	27
Cetus	Ceti	Cet	4
Chamaeleon	Chamaeleontis	Cha	79
Circinus	Circini	Cir	85
Columba	Columbae	Col	54
Coma Berenices	Comae Berenices	Com	42
Corona Australis	Coronae Australis	CrA	80
Corona Borealis	Coronae Borealis	CrB	73
Corvus	Corvi	Crv	70
Crater	Crateris	Crt	53
Crux	Crucis	Cru	88
Cygnus	Cygni	Cyg	16
Delphinus	Delphini	Del	69
Dorado	Doradus	Dor	72
Draco	Draconis	Dra	8

Name	Genitive	Abbreviation	Order of Size
Equuleus	Equulei	Equ	87
Eridanus	Eridani	Eri	6
Fornax	Fornacis	For	41
Gemini	Geminorum	Gem	30
Grus	Gruis	Gru	45
Hercules	Herculis	Her	5
Horologium	Horologii	Hor	58
Hydra	Hydrae	Hya	1
Hydrus	Hydri	Hyi	61
Indus	Indi	Ind	49
Lacerta	Lacertae	Lac	68
Leo	Leonis	Leo	12
Leo Minor	Leonis Minoris	LMi	64
Lepus	Leporis	Lep	51
Libra	Librae	Lib	29
Lupus	Lupi	Lup	46
Lynx	Lyncis	Lyn	28
Lyra	Lyrae	Lyr	52
Mensa	Mensae	Men	75
Microsopium	Microscopii	Mic	66
Monoceros	Monocerotis	Mon	35
Musca	Muscae	Mus	77
Norma	Normae	Nor	74
Octans	Octantis	Oct	50
Ophiuchus	Ophiuchi	Oph	11
Orion	Orionis	Ori	26
Pavo	Pavonis	Pav	44
Pegasus	Pegasi	Peg	7
Perseus	Persei	Per	24
Phoenix	Phoenicis	Phe	37
Pictor	Pictoris	Pic	59
Pisces	Piscium	Psc	14
Piscis Austrinus	Piscis Austrini	PsA	60
Puppis	Puppis	Pup	20
Pyxis	Pyxidis	Pyx	65
Reticulum	Reticuli	Ret	82
Sagitta	Sagittae	Sge	86
Sagittarius	Sagittarii	Sgr	15
Scorpius	Scorpii	Sco	33
Sculptor	Sculptoris	Scl	36
Scutum	Scuti	Sct	84

(*continued*)

Name	Genitive	Abbreviation	Order of Size
Serpens	Serpentis	Ser	23
Sextans	Sextantis	Sex	47
Taurus	Tauri	Tau	17
Telescopium	Telescopii	Tel	57
Triangulum	Trianguli	Tri	78
Triangulum Australe	Trianguli Australis	TrA	83
Tucana	Tucanae	Tuc	48
Ursa Major	Ursae Majoris	UMa	3
Ursa Minor	Ursae Minoris	UMi	56
Vela	Velorum	Vel	32
Virgo	Virginis	Vir	2
Volans	Volantis	Vol	76
Vulpecula	Vulpeculae	Vul	55

✦ APPENDIX 3 ✦

SOLAR ECLIPSES, 1998–2005

Date and Time (UT)	Type	Max. Duration (min., sec.)[a]	Area of visibility[b]
1998 Feb. 26 17	Total	4:08	NW South America, S. Caribbean
1998 Aug. 22 2	Annular	3:14	Indonesia, Polynesia
1999 Feb. 16 7	Annular	1:19	Australia
1999 Aug. 11 11	Total	2:23	Europe, SW and S Asia
2000 Feb. 5 13	Partial (0.58)[c]	——	Antarctica
2000 Jul. 1 20	Partial (0.48)[c]	——	S of S America
2000 Jul. 31 2	Partial (0.60)[c]	——	N Asia, NW of N America, Arctic
2000 Dec. 25 18	Partial (0.72)[c]	——	All N America except Alaska
2001 Jun. 21 12	Total	4:56	S Africa, Madagascar
2001 Dec. 14 21	Annular	3:54	Central America
2002 Jun. 10 24	Annular	1:13	Pacific Ocean
2002 Dec. 4 8	Total	2:04	S Africa, Australia
2003 May 31 4	Annular	3:37	Iceland
2003 Nov. 23 23	Total	1:57	Antarctica
2004 Apr. 19 14	Partial (0.74)[c]	——	S of Africa, parts of Antarctica
2004 Oct. 14 3	Partial (0.93)[c]	——	NE Asia, Hawaii, parts of Alaska
2005 Apr. 8 21	Annular-Total	0:42	Central America, N. of S America
2005 Oct. 3 11	Annular	4:32	Spain, Africa

[a] Maximum duration at the place on the eclipse track where totality or annularity is longest.

[b] For total and annular eclipses, only parts of these regions get to see the total or annular eclipse; the rest experience only a partial eclipse.

[c] Maximum magnitude: largest fraction of way across Sun that Moon extends.

✦ APPENDIX 4 ✦

TOTAL AND PARTIAL LUNAR ECLIPSES, 1998–2005

If the Universal Time of maximum eclipse occurs during the nighttime hours at your location, the eclipse will be visible from your location (weather permitting).

Date	Time Max. Ecl. (UT)	Magnitude Max. Ecl.[a]
1999 Jul. 28	11:32	0.40
2000 Jan. 21	4:43	1.33
2000 Jul. 16	13:55	1.77
2001 Jan. 9	20:20	1.19
2001 Jul. 5	14:55	0.49
2003 May 16	3:39	1.13
2003 Nov. 9	1:18	1.02
2004 May 4	20:30	1.30
2004 Oct. 28	3:03	1.31
2005 Oct. 17	12:02	0.06

Note: There are no total or partial lunar eclipses in 1998 or 2002.

[a] Magnitude is the fraction of the Moon's diameter which is within the umbra (thus, an eclipse magnitude of more than 1.00 means the eclipse is total, less than 1.00 is partial).

✦ APPENDIX 5 ✦

OBSERVATIONAL DATA FOR THE PLANETS

Planet	Min. Diam.[a]	Max. Diam.[b]	Mean Opp. Diam.[c]	Mean Opp. Mag.[d]
Mercury	4.5	13.0	7.8[e]	0.0[e]
Venus	9.6	65.4	25.2[e]	−4.4[e]
Mars	3.5	25.7	17.9	−2.0
Jupiter	30.4	50.0	46.8	−2.7
Saturn	14.9	20.7	19.4	−0.2[f]
Uranus	3.3	4.1	3.9	+5.5
Neptune	2.1	2.3	2.3	+7.8
Pluto	—	—	—	+14

[a] Angular diameter in arc-seconds at greatest distance.

[b] Angular diameter in arc-seconds at least distance.

[c] Angular diameter in arc-seconds at mean opposition distance.

[d] Visual magnitude at mean opposition distance.

[e] At greatest elongation, not opposition.

[f] With rings fully opened; magnitude +0.7 with rings closed.

✦ APPENDIX 6 ✦

PLANETARY ORBITAL DATA
AND SYNODIC PERIODS

Planet	Mean Dist. (AU)	Sid. Per.[a]	Mean Syn. Per.[b]
Mercury	0.387	87.969d	115.88d
Venus	0.723	224.701d	583.92d
Earth	1.000	365.256d	—
Mars	1.524	686.980d	779.94d
Jupiter	5.203	11.862y	398.88d
Saturn	9.539	29.457y	378.09d
Uranus	19.182	84.010y	369.66d
Neptune	30.058	164.793y	367.49d
Pluto	39.44	248.5y	366.73d

[a] Sidereal orbital period.
[b] Mean synodic period.

◆ APPENDIX 7 ◆

PHYSICAL DATA FOR THE PLANETS

Planet	Eq. Diam. (km)[a]	Mass	Volume	Mean Dens.[b]	Alb.[c]
Mercury	4,878	0.06	0.06	5.43	0.11
Venus	12,104	0.82	0.86	5.24	0.65
Earth	12,756	1.00	1.00	5.52	0.37
Mars	6,787	0.11	0.15	3.94	0.15
Jupiter	142,800	317.83	1,323	1.33	0.52
Saturn	120,000	95.16	752	0.70	0.47
Uranus	50,800	14.50	64	1.30	0.51
Neptune	48,600	17.20	54	1.76	0.41
Pluto	2,250	0.002	0.01	c.2	c.0.3

[a] Equatorial diameter.
[b] Mean density (density of water = 1).
[c] Albedo (0.0 = completely nonreflective; 1.0 = completely reflective).

✦ APPENDIX 8 ✦

IMPORTANT POSITIONS OF
THE INFERIOR PLANETS, 1998–2005

Greatest Elongations of Mercury

Mercury is best seen from two weeks before until a single week after an evening greatest elongation, and one week before until two weeks after a morning greatest elongation. Mercury is far more favorably seen at evening elongations that occur within a few months of spring's start and morning elongations that occur within a few months of autumn's start. Below, M = morning and E = evening.

During the period 1998–2005, transits of Mercury occur on Nov. 15, 1999, and May 7, 2003 (but neither is visible from the United States).

1998: Jan. 6 (M), Mar. 20 (E), May 4 (M), Jul. 17 (E), Aug. 31 (M), Nov. 11 (E), Dec. 20 (M)
1999: Mar. 3 (E), Apr. 16 (M), Jun. 28 (E), Aug. 14 (M), Oct. 24 (E), Dec. 3 (M)
2000: Feb. 15 (E), Mar. 28 (M), Jun. 9 (E), Jul. 27 (M), Oct. 6 (E), Nov. 15 (M)
2001: Jan. 28 (E), Mar. 11 (M), May 22 (E), Jul. 9 (M), Sep. 18 (E), Oct. 29 (M)
2002: Jan. 11 (E), Feb. 21 (M), May 4 (E), Jun. 21 (M), Sep. 1 (E), Oct. 13 (M), Dec. 26 (E)
2003: Feb. 4 (M), Apr. 16 (E), Jun. 3 (M), Aug. 14 (E), Sep. 27 (M), Dec. 9 (E)
2004: Jan. 17 (M), Mar. 29 (E), May 14 (M), Jul. 27 (E), Sep. 9 (W), Nov. 21 (E), Dec. 29 (M)
2005: Mar. 12 (E), Apr. 26 (M), Jul. 9 (E), Aug. 23 (W), Nov. 3 (E), Dec. 12 (W)

Venus

Sup. Conj.	Gr. Eve Elong.	Inf. Conj.	Gr. Morn Elong.
		1998 Jan. 16	1998 Mar. 27
1998 Oct. 20	1999 Jun. 11	1999 Aug. 20	1999 Oct. 30
2000 Jun. 11	2001 Jan. 17	2001 Mar. 30	2001 Jun. 8
2002 Jan. 14	2002 Aug. 22	2002 Oct. 31	2003 Jan. 11
2003 Aug. 18	2004 Mar. 29	2004 Jun. 8[a]	2004 Aug. 17
2005 Mar. 31	2005 Nov. 3		

[a] Transit of Venus.

✦ APPENDIX 9 ✦

OPPOSITIONS AND CONJUNCTIONS OF THE SUPERIOR PLANETS, 1998–2005

Oppositions

	Mars	Jupiter	Saturn	Uranus	Neptune	Pluto
1998		Sep. 16	Oct. 23	Aug. 3	Jul. 23	May 28
1999	Apr. 24	Oct. 23	Nov. 6	Aug. 7	Jul. 26	May 31
2000		Nov. 28	Nov. 19	Aug. 11	Jul. 27	Jun. 1
2001	Jun. 13		Dec. 3	Aug. 15	Jul. 30	Jun. 4
2002		Jan. 1	Dec. 17	Aug. 20	Aug. 2	Jun. 7
2003	Aug. 28	Feb. 2	Dec. 31	Aug. 24	Aug. 4	Jun. 9
2004		Mar. 4		Aug. 27	Aug. 6	Jun. 11
2005	Nov. 7	Apr. 3	Jan. 13	Sep. 1	Aug. 8	Jun. 14

Conjunctions with Sun

	Mars	Jupiter	Saturn	Uranus	Neptune	Pluto
1998	May 12	Feb. 23	Apr. 13	Jan. 28	Jan. 19	Nov. 30
1999		Apr. 1	Apr. 27	Feb. 2	Jan. 22	Dec. 3
2000	Jul. 1	May 8	May 10	Feb. 6	Jan. 24	Dec. 4
2001		Jun. 14	May 25	Feb. 9	Jan. 26	Dec. 7
2002	Aug. 10	Jul. 20	Jun. 9	Feb. 13	Jan. 28	Dec. 9
2003		Aug. 22	Jun. 24	Feb. 17	Jan. 30	Dec. 12
2004	Sep. 15	Sep. 21	Jul. 8	Feb. 22	Feb. 2	Dec. 13
2005		Oct. 22	Jul. 23	Feb. 25	Feb. 3	Dec. 16

✦ APPENDIX 10 ✦

POSITIONS OF MARS, JUPITER, AND SATURN, 1998–2003

Positions (R.A. in hrs. and mins., then declination in degrees) for 0 UT of first day in the month mentioned. Locate these positions on the equatorial star map in this book (see Figure 21) or any sky map with R.A. and declination marked on it.

Mars

	1998		1999		2000		2001		2002		2003	
	R.A.	Dec	R.A.	Dec	R.A.	Dec	R.A.	Dec	R.A.	Dec	R.A.	Dec
Jan.	20 54	−18.7	13 11	−5.4	22 01	−13.3	14 12	−12.0	23 13	−5.8	15 09	−17.0
Feb.	22 30	−10.5	14 03	−10.3	23 30	−4.0	15 22	−17.3	0 35	3.6	16 32	−21.5
Mar.	23 51	−1.8	14 35	−12.9	5 01	5.1	16 22	−20.7	1 50	11.5	17 49	−23.4
Apr.	1 18	7.8	14 37	−13.2	2 17	13.8	17 14	−22.8	3 15	18.6	19 15	−23.0
May	2 43	15.7	14 00	−11.0	3 43	20.1	17 53	−24.1	4 40	23.0	20 32	−20.4
Jun.	4 14	21.5	13 31	−9.7	5 14	23.7	17 44	−25.9	6 10	24.4	21 43	−16.6
Jul.	5 44	23.9	13 45	−12.0	6 42	24.0	17 04	−26.8	7 35	22.7	22 35	−13.6
Aug.	7 15	23.2	14 34	−16.7	8 10	21.2	16 58	−26.9	8 58	18.4	22 56	−13.4
Sep.	8 39	19.5	15 45	−21.7	9 31	16.0	17 43	−27.0	10 16	12.0	22 34	−16.0
Oct.	9 55	14.1	17 10	−24.8	10 44	9.4	18 56	−25.7	11 27	4.7	22 16	−15.7
Nov.	11 06	7.4	18 48	−24.7	11 15	1.9	20 23	−21.6	12 40	−3.2	22 40	−11.0
Dec.	12 10	0.8	20 25	−20.7	13 03	−5.3	21 48	−14.7	13 51	−10.5	23 29	−4.3

Jupiter

	1998		1999		2000		2001		2002		2003	
	R.A.	Dec	R.A.	Dec	R.A.	Dec	R.A.	Dec	R.A.	Dec	R.A.	Dec
Jan.	21 40	−14.9	23 32	−4.3	1 35	8.6	4 01	19.8	6 46	23.0	9 18	16.5
Apr.	22 59	−7.5	0 42	3.4	2 28	13.7	4 24	21.1	6 31	23.4	8 43	19.1
Jul.	23 53	−2.1	1 55	10.5	3 53	19.4	5 48	23.1	7 39	21.8	9 23	16.2
Oct.	23 30	−4.9	2 05	11.0	4 39	21.2	7 01	22.5	8 59	17.6	10 38	9.6

Saturn

	1998 R.A.	1998 Dec	1999 R.A.	1999 Dec	2000 R.A.	2000 Dec	2001 R.A.	2001 Dec	2002 R.A.	2002 Dec	2003 R.A.	2003 Dec
Jan.	0 55	3.1	1 43	7.9	2 35	12.6	3 31	16.8	4 32	20.1	5 36	22.0
Apr.	1 24	6.4	2 08	10.6	2 55	14.5	3 44	17.9	4 36	20.6	5 32	22.3
Jul.	2 02	9.9	2 50	13.9	3 39	17.9	4 30	20.1	5 22	21.9	6 15	22.6
Oct.	2 03	9.6	2 58	14.2	3 56	18.1	4 56	20.8	5 56	22.1	6 55	22.1

◆ APPENDIX 11 ◆

THE MESSIER OBJECTS

M	NGC	h	m	°	′	Constellation	Size[a] ′	Integrated Magnitude	Description
1[b]	1952	05	34.5	+22	01	Tau	6 × 4	c. 8.4	Supernova remnant
2	7089	21	33.5	−00	49	Aqr	13	6.5	Globular cluster
3	5272	13	42.2	+28	23	CVn	16	6.4	Globular cluster
4	6121	16	23.6	−26	32	Sco	26	5.9	Globular cluster
5	5904	15	18.6	+02	05	Ser	17	5.8	Globular cluster
6	6405	17	40.1	−32	13	Sco	15	4.2	Open cluster
7	6475	17	53.9	−34	49	Sco	80	3.3	Open cluster
8[b]	6523	18	03.8	−24	23	Sgr	90 × 40	c. 5.8	Diffuse nebula
9	6333	17	19.2	−18	31	Oph	9	c. 7.9	Globular cluster
10	6254	16	57.1	−04	06	Oph	15	6.6	Globular cluster
11[b]	6705	18	51.1	−06	16	Sct	14	5.8	Open cluster
12	6218	16	47.2	−01	57	Oph	14	6.6	Globular cluster
13	6205	16	41.7	+36	28	Her	17	5.9	Globular cluster
14	6402	17	37.6	−03	15	Oph	12	7.6	Globular cluster
15	7078	21	30.0	+12	10	Peg	12	6.4	Globular cluster
16[b]	6611	18	18.8	−13	47	Ser	7	6.0	Open cluster
17[b]	6618	18	20.8	−16	11	Sgr	46 × 37	7	Diffuse nebula
18	6613	18	19.9	−17	08	Sgr	9	6.9	Open cluster
19	6273	17	02.6	−26	16	Oph	14	7.2	Globular cluster
20[b]	6514	18	02.6	−23	02	Sgr	29 × 27	c. 8.5	Diffuse nebula
21	6531	18	04.6	−22	30	Sgr	13	5.9	Open cluster
22	6656	18	36.4	−23	54	Sgr	24	5.1	Globular cluster
23	6494	17	56.8	−19	01	Sgr	27	5.5	Open cluster
24[b]		18	16.9	−18	29	Sgr	90	c. 4.5	
25	IC 4725	18	31.6	−19	15	Sgr	32	4.6	Open cluster
26	6694	18	45.2	−09	24	Sct	15	8.0	Open cluster
27[b]	6853	19	59.6	+22	43	Vul	8 × 4	c. 8.1	Planetary nebula
28	6626	18	24.5	−24	52	Sgr	11	c. 6.9	Globular cluster
29	6913	20	23.9	+38	32	Cyg	7	6.6	Open cluster
30	7099	21	40.4	−23	11	Cap	11	7.5	Globular cluster
31[b]	224	00	42.7	+41	16	And	178 × 63	3.4	Spiral galaxy
32	221	00	42.7	+40	52	And	8 × 6	8.2	Elliptical galaxy
33	598	01	33.9	+30	39	Tri	62 × 39	5.7	Spiral galaxy

| Number | | RA 2000.0 Dec. | | | | Constellation | Size[a] | Integrated | Description |
M	NGC	h	m	°	′		′	Magnitude	
34	1039	02	42.0	+42	47	Per	35	5.2	Open cluster
35	2168	06	08.9	+24	20	Gem	28	5.1	Open cluster
36	1960	05	36.1	+34	08	Aur	12	6.0	Open cluster
37	2099	05	52.4	+32	33	Aur	24	5.6	Open cluster
38	1912	05	28.7	+35	50	Aur	21	6.4	Open cluster
39	7092	21	32.2	+48	26	Cyg	32	4.6	Open cluster
40[b]		12	22.4	+58	05	UMa		8	
41	2287	06	47.0	−20	44	CMa	38	4.5	Open cluster
42[b]	1976	05	35.4	−05	27	Ori	66 × 60	4	Diffuse nebula
43[b]	1982	05	35.6	−05	16	Ori	20 × 15	9	Diffuse nebula
44[b]	2632	08	40.1	+19	59	Cnc	95	3.1	Open cluster
45[b]		03	47.0	+24	07	Tau	110	1.2	Open cluster
46	2437	07	41.8	−14	49	Pup	27	6.1	Open cluster
47	2422	07	36.6	−14	30	Pup	30	4.4	Open cluster
48	2548	08	13.8	−05	48	Hya	54	5.8	Open cluster
49	4472	12	29.8	+08	00	Vir	9 × 7	8.4	Elliptical galaxy
50	2323	07	03.2	−08	20	Mon	16	5.9	Open cluster
51[b]	5194-5	13	29.9	+47	12	CVn	11 × 8	8.1	Spiral galaxy
52	7654	23	24.2	+61	35	Cas	13	6.9	Open cluster
53	5024	13	12.9	+18	10	Com	13	7.7	Globular cluster
54	6715	18	55.1	−30	29	Sgr	9	7.7	Globular cluster
55	6809	19	40.0	−30	58	Sgr	19	7.0	Globular cluster
56	6779	19	16.6	−30	11	Lyr	7	8.2	Globular cluster
57[b]	6720	18	53.6	+33	02	Lyr	1	c. 9.0	Planetary nebula
58	4579	12	37.7	+11	49	Vir	5 × 4	9.8	Spiral galaxy
59	4621	12	42.0	+11	39	Vir	5 × 3	9.8	Elliptical galaxy
60	4649	12	43.7	+11	33	Vir	7 × 6	8.8	Elliptical galaxy
61	4303	12	21.9	+04	28	Vir	6 × 5	9.7	Spiral galaxy
62	6266	17	01.2	−30	07	Oph	14	6.6	Globular cluster
63	5055	13	15.8	+42	02	CVn	12 × 8	8.6	Spiral galaxy
64[b]	4826	12	56.7	+21	41	Com	9 × 5	8.5	Spiral galaxy
65	3623	11	18.9	+13	05	Leo	10 × 3	9.3	Spiral galaxy
66	3627	11	20.2	+12	59	Leo	9 × 4	9.0	Spiral galaxy
67	2682	08	50.4	+11	49	Cnc	30	6.9	Open cluster
68	4590	12	39.5	−26	45	Hya	12	8.2	Globular cluster
69	6637	18	31.4	−32	21	Sgr	7	7.7	Globular cluster
70	6681	18	43.2	−32	18	Sgr	8	8.1	Globular cluster
71	6838	19	53.8	+18	47	Sge	7	8.3	Globular cluster
72	6981	20	53.5	−12	32	Aqr	6	9.4	Globular cluster
73[b]	6994	20	58.9	−12	38	Aqr			

(continued)

| Number | | RA 2000.0 Dec. | | | | Constellation | Size[a] | Integrated Magnitude | Description |
M	NGC	h	m	°	′		′		
74	628	01	36.7	+15	47	Psc	10 × 9	9.2	Spiral galaxy
75	6864	20	06.1	−21	55	Sgr	6	8.6	Globular cluster
76	650-1	01	42.4	+51	34	Per	2 × 1	c. 11.5	Planetary nebula
77	1068	02	42.7	−00	01	Cet	7 × 6	8.8	Spiral galaxy
78	2068	05	46.7	+00	03	Ori	8 × 6	8	Diffuse Nebula
79	1904	05	24.5	−24	33	Lep	9	8.0	Globular cluster
80	6093	16	17.0	−22	59	Sco	9	7.2	Globular cluster
81	3031	09	55.6	+69	04	UMa	26 × 14	6.8	Spiral galaxy
82	3034	09	55.8	+69	41	UMa	11 × 5	8.4	Irregular galaxy
83	5236	13	37.0	−29	52	Hya	11 × 10	c. 7.6	Spiral galaxy
84	4374	12	25.1	+12	53	Vir	5 × 4	9.3	Elliptical galaxy
85	4382	12	25.4	+18	11	Com	7 × 5	9.2	Elliptical galaxy
86	4406	12	26.2	+12	57	Vir	7 × 6	9.2	Elliptical galaxy
87	4486	12	30.8	+12	24	Vir	7	8.6	Elliptical galaxy
88	4501	12	32.0	+14	25	Com	7 × 4	9.5	Spiral galaxy
89	4552	12	35.7	+12	33	Vir	4	9.8	Elliptical galaxy
90	4569	12	36.8	+13	10	Vir	10 × 5	9.5	Spiral galaxy
91	4548	12	35.4	+14	30	Com	5 × 4	10.2	Spiral galaxy
92	6341	17	17.1	+43	08	Her	11	6.5	Globular cluster
93	2447	07	44.6	−23	52	Pup	22	c. 6.2	Open cluster
94	4736	12	50.9	+41	07	CVn	11 × 9	8.1	Spiral galaxy
95	3351	10	44.0	+11	42	Leo	7 × 5	9.7	Spiral galaxy
96	3368	10	46.8	+11	49	Leo	7 × 5	9.2	Spiral galaxy
97[b]	3587	11	14.8	+55	01	UMa	3	c. 11.2	Planetary nebula
98	4192	12	13.8	+14	54	Com	10 × 3	10.1	Spiral galaxy
99	4254	12	18.8	+14	25	Com	5	9.8	Spiral galaxy
100	4321	12	22.9	+15	49	Com	7 × 6	9.4	Spiral galaxy
101	5457	14	03.2	+54	21	UMa	27 × 26	7.7	Spiral galaxy
102[b]									
103	581	01	33.2	+60	42	Cas	6	c. 7.4	Open cluster
104[b]	4594	12	40.0	−11	37	Vir	9 × 4	8.3	Spiral galaxy
105	3379	10	47.8	+12	35	Leo	4 × 4	9.3	Elliptical galaxy
106	4258	12	19.0	+47	18	CVn	18 × 8	8.3	Spiral galaxy
107	6171	16	32.5	−13	03	Oph	10	8.1	Globular cluster
108	3556	11	11.5	+55	40	UMa	8 × 2	10.0	Spiral galaxy
109	3992	11	57.6	+53	23	UMa	8 × 5	9.8	Spiral galaxy
110	205	00	40.4	+41	41	And	17 × 10	8.0	Elliptical galaxy

Source: A. Hirshfeld and R. W. Sinnott, eds., *Sky Catalogue 2000.0,* Vol. 2 (Sky Publishing Corp./Cambridge University Press, 1985).

[a] The dimensions given are as seen on long-exposure photographs and, for galaxies in particular, are larger than the sizes that will be seen visually.

[b]M1	Crab Nebula
M8	Lagoon Nebula; contains a star cluster
M 11	Wild Duck Cluster
M 16	Surrounded by the Eagle Nebula
M 17	Omega Nebula
M 20	Trifid Nebula
M 24	Star field in Sagittarius, containing the open cluster NGC 6603
M 27	Dumbbell Nebula
M 31	Andromeda Galaxy
M 40	Faint double star Winnecke 4, mags. 9.0 and 9.6
M 42, M 43	Orion Nebula
M 44	Praesepe, the Beehive Cluster
M 45	The Pleiades; no NGC or IC number
M 51	Whirlpool Galaxy
M 57	Ring Nebula
M 64	Black-Eye Galaxy
M 73	Small group of four faint stars
M 97	Owl Nebula
M 102	Duplicate of M 101
M 104	Sombrero Galaxy

✦ APPENDIX 12 ✦

SOME BASICS OF
BINOCULARS AND TELESCOPES

BEFORE BUYING YOUR FIRST TELESCOPE

A good first telescope is an investment of at least hundreds of dollars, and of much hope and anticipation. That is why it is so unfortunate that many beginning skywatchers buy a telescope before they are ready and end up becoming so frustrated that they give up on astronomy altogether.

I advise beginners to ask themselves a series of questions to determine if they are really ready to purchase their first telescope. Do you know at least some of the major constellations? Can you recognize and locate most of the major planets? Have you read a few books on astronomy, preferably at least one guide to telescope use (or at least one detailed chapter on the subject)? Have you considered where you are going to keep your telescope and where you will use it? Will you be able to carry it easily from one place to another? Have you spent enough time under the night sky to determine you are comfortable outdoors in the dark and cold? Have you practiced naked-eye observation and found that there is much to enjoy in it for its own sake?

The final question to ask yourself before you purchase a telescope is whether or not you've practiced astronomy with a marvelous optical instrument that is generally much cheaper, more portable, and easier to use than a telescope. In other words, have you ever tried astronomy with a good pair of binoculars?

BINOCULARS

Binoculars are not inferior to telescopes; they complement them. A basic pair of binoculars can reveal hundreds of features on the Moon and show dozens of times more stars, star clusters, and other deep-sky objects than the naked eye can make out on its own. When there is a conjunction of planets or a long comet tail, binoculars may provide the best view of them.

There is a lot to learn about the selection, care, and maintenance of binoculars, and several good books are devoted to this subject. One great all-around work on binocular astronomy is Phil Harrington's *Touring the Universe Through Binoculars* (see Sources of Information). A serviceable pair of binoculars costs so little—or may be so easily borrowed from friends or family members—that detailed knowledge of them is not necessary for the novice. The basic points will get you started.

The key numbers you see on a pair of binoculars are its magnification and the diameter of its primary (large) lenses expressed in millimeters. Thus "7 × 50" binoculars have a magnification of 7× (they make things look seven times larger than they appear to the naked eye) and their primary lenses are 50mm in diameter (about two inches wide).

As with telescopes, the most important property of binoculars for astronomy is their light-gathering ability. This is determined by the size of the primary lenses. For example, 7 × 50 binoculars would usually be preferable to 7 × 35s for astronomy. The latter typically weigh less and might therefore be the choice for activities such as birding in broad daylight. Most of the objects in the heavens, however, are faint, and their viewing benefits tremendously from the aid of instruments with superior light grasp.

Are the 10 × 80, 12 × 80, and 20 × 80 "giant binoculars" you may see advertised in astronomy magazines and product catalogues an even better purchase for beginning astronomers? No. These instruments are simply too heavy to hold, and they have narrower fields of vision than the 7 × 50s. They are also many times more expensive. They are a wonderful acquisition for the more experienced observer who already owns 7 × 50s or similar binoculars and a telescope or two, but they are not a good choice as a first pair of binoculars.

Even 10 × 50 binoculars tend to have too narrow a field of view to keep really steady on a celestial object. Such binoculars should be used with a tripod rather than held in the hands.

Basic Parts and Types of Telescope

Magnification may not be quite as important overall in astronomy as light-gathering power, but of course the great limitation of binoculars is that they don't provide high enough magnification to see many of the finest sights in the universe. To see the rings of Saturn and the cloud bands of Jupiter properly, or to study the intricate interiors of lunar craters and the details of a galaxy or nebula, you must have the magnification that only telescopes can supply.

All telescopes feature a primary lens or mirror. The wider this is, the greater the telescope's light-gathering power and its "resolution"—its ability to show detail. A variety of magnifications are supplied by **oculars**, also known as **eyepieces**. A small telescope can be used with a high-power eyepiece to obtain very high magnification, but the image will be so fuzzy that it is useless. The rule of thumb is that for every inch of **aperture** (that is, diameter of the primary lens or mirror), you can use no more than about 50× and still get a usefully sharp image. In other words, if you have a six-inch telescope (six-inch-wide primary mirror, let's say), you can use a maximum of $50 \times 6 = 300\times$. And that power will only be usable if the "seeing" (atmospheric steadiness) is very good and if you use the telescope on certain kinds of celestial sights like double stars and some planetary details. You'll usually be better off with lower magnifications when viewing extended, fuzzy, and perhaps low-surface-brightness objects like nebulae.

As already noted, the primary or "objective" in your telescope may be either a lens or a mirror. The **refractor telescope** uses a primary lens that, with the help of additional lenses, focuses the light of celestial objects to your eyepiece. This is the kind of telescope best known to the general public, the kind with its primary lens on its skyward end. More popular among amateur astronomers is the **reflector telescope**, which typically has a primary mirror, instead of a lens, and is focused to a secondary mirror at the telescope's skyward end. The secondary mirror directs the image of the celestial object to the eyepiece. The reflector telescope has become more popular than the refractor because it can be much less expensive and smaller per inch of aperture. The difference in size and cost is most marked with telescopes with aperture sizes above about three inches. A basic *Newtonian reflector* with an eight-inch aperture can cost as little as a few hundred dollars with certain kinds of mounts. However, many amateur astronomers looking for an even more compact

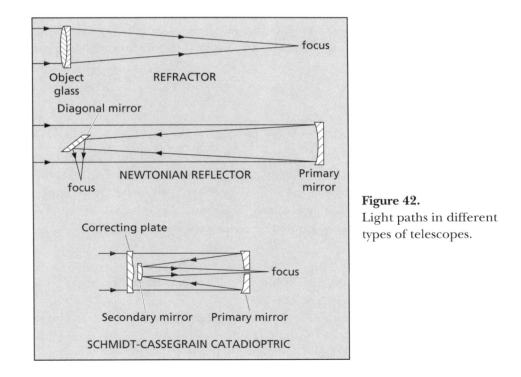

Figure 42.
Light paths in different types of telescopes.

telescope have purchased, usually for a few thousand dollars, a third major type of telescope, the **catadioptric**. A catadioptric telescope, such as the *Schmidt-Cassegrain,* uses a combination of lens and mirror in a folded light path.

Every telescope should come with a small "finderscope," whose wider field of view will help locate the target object so that you can get it into the field of the main telescope. Every telescope should also come with a sturdy mount, steady enough so that when the telescope tube is moved, the image in the eyepiece does not shake for more than a second at most.

There are several different kinds of mounts for telescopes. An **equatorial mount** is designed to have one axis pointed to the north celestial pole and to permit motion of the telescope's tube so that a celestial object can be followed with one movement as the Earth's rotation causes it to progress across the sky. This kind of mount, equipped with a "clock drive" to keep the celestial object centered, is generally necessary for long-exposure astrophotography. There are some advantages to having an **altazimuth mount**—a mount that lets you move the telescope tube in altitude and azimuth rather than right ascension and declination as an equatorial mount does. One advantage of an altazimuth mount is its lighter weight. This is especially true with the type of altazimuth mount called a **Dobsonian mount**. The simple and inexpensive structure of the Dobsonian has made affordable and portable telescopes of far greater aperture than were previously available to amateur astronomers.

The alignment of lenses or mirrors in an optical system is known as **collimation**. Refractors and catadioptric telescopes are generally designed to stay in collimation almost indefinitely, as long as they are used properly. Reflectors typically slip out of collimation more often and have to be adjusted by the user. Instructions on how to adjust the primary and secondary mirrors to obtain collimation are provided with new telescopes.

GLOSSARY

✦　✦　✦

altazimuth system: System for indicating positions in the sky using altitude and azimuth as vertical and horizontal measure.

altitude: Apparent angular height in the sky (vertical measurement in the altazimuth system).

angular altitude: Expressed in degrees (°), minutes ('), and seconds (") of arc. The moon is about ½° or 30' wide; Venus is never more than about 1' or 60" wide.

annular eclipse: Solar eclipse in which the entire Moon passes in front of the Sun but is too far out in its orbit to cover the Sun completely, thus leaving a ring (Latin "annulus") of exposed solar surface.

aphelion: The far point in an orbit around the Sun.

apogee: The far point in an orbit around the Earth.

apparition: Period of a planet's (or other object's) visibility between two spells when it is not viewable.

appulse: The closest apparent approach in the sky (not in space) between two celestial objects.

artificial satellite: (see "satellite")

asterism: A pattern of stars in the sky that is not an official constellation.

asteroid: A small rocky world of the class found mostly between the orbits of Mars and Jupiter (also called a "minor planet").

atmospheric extinction: Dimming of the light of celestial objects due to absorption and scattering by Earth's atmosphere.

aurora: Patterns of radiance (usually moving and fluctuating) produced when atomic particles from the Sun are energized in Earth's magnetic field and channeled to collide with upper-atmosphere gases in regions surrounding Earth's magnetic poles. Also known as "the Northern Lights" or "aurora borealis" ("aurora australis" in the Southern Hemisphere).

averted vision: Technique of looking slightly to the side of a faint object in order to increase its visibility by allowing its light to fall on the parts of the eye's retina most sensitive to light.

azimuth: Horizontal measure around the sky in the altazimuth system.

Bailey's beads: Small segments of the Sun's blinding surface shining through lowlands on the Moon's edge just before the start or just after the end of totality in a total solar eclipse.

belts: The dark bands of clouds on the faces of Jupiter and Saturn.

binary star: A double star system in which the members are believed to be orbiting each other (that is, around a common center of gravity).

black hole: An object thought to be the result of a massive star's collapse, whose gravity has become so strong that it prevents even light (and other electromagnetic radiations) from escaping.

blue giant: A massive, large, and extremely luminous star with a very high surface temperature that causes it to appear bluish-white to the eye.

Bobrovnikoff method: A method for estimating the brightness of a comet, in which the comet's coma is defocused until it appears more similar to the size of defocused stars, to whose known brightness it can then be compared.

celestial objects: Bodies in outer space beyond Earth's atmosphere.

celestial sphere: The imaginary sphere surrounding Earth, whose inner surface is the entire sky both above and below one's horizon.

Cepheid: A type of pulsating variable star whose precisely regular brightness variations can be employed to estimate the star's distance by use of the "period-luminosity relation" (if you know the period of the Cepheid's brightness variations, you know its luminosity—true brightness—and can therefore compare this to its apparent brightness to determine distance).

circumpolar: Close enough to one of the celestial poles so as to never rise or set but rather circle around the pole above the horizon.

coma: The cloud of gas and dust released from a comet's "nucleus," which, together with the tremendously smaller solid "nucleus," forms the "head" of the comet.

comet: A mostly icy body, tremendously less massive than the planets, that produces a cloud of dust and/or gas ("coma") when it is heated sufficiently by the Sun (or, less commonly, by other forces).

companion star: (see "double star").

conjunction: Strictly speaking, the arrangement when one celestial object moves to have the same right ascension or "ecliptic longitude" (passing due north or due south of a second object in a celestial coordinate system) of another. More loosely, any temporary pairing or gathering of celestial objects.

constellation: An official pattern of stars or, more strictly, the officially demarcated section of sky in which that pattern lies.

corona (solar corona): The outer atmosphere of the Sun, which can be seen as a pearly glow when the blindingly bright surface of the Sun is completely hidden during a total eclipse of the Sun.

crescent: Phase in which a world's hemisphere that is facing us appears less than half lit.

culminate: To reach the north-south meridian of the sky, typically achieving the highest point above horizon.

Danjon scale: A scale of verbal descriptions for estimating the brightness of a total lunar eclipse.

dark adaptation: The increase in the sensitivity of our eyes to dim light that occurs when they are kept away from bright light for a while.

dark nebula: A nebula that does not shine by either emitted or reflected light and is therefore visible only in silhouette against a more distant bright nebula or starry background.

D.C.: (see "degree of condensation").

declination: North-south measure in the equatorial system of celestial coordinates, corresponding to latitude on Earth.

deep-sky object: An object beyond our solar system, though the term is usually not applied to individual stars or double and multiple star systems but rather to star clusters, nebulae, and galaxies.

degree of condensation (D.C.): The degree to which the light of a comet is concentrated toward the center.

diamond-ring effect: The appearance at the start and end of some total solar eclipses of a first (or last) starlike point of the Sun's surface, which resembles a diamond on the band of the still-visible solar corona when seen through a valley on the Moon's edge.

dichotomy: A world's appearance of being exactly half-lit (it does not happen precisely at the time it theoretically should for Venus and Mercury).

diffuse nebula: A luminous nebula that shines from reflecting the light of nearby stars ("reflection nebula") or is heated enough by very hot stars to glow on its own ("emission nebula").

direct motion: The more typical eastward motion of planets with respect to the background of stars.

diurnal motion: The apparent motion of celestial objects caused by the Earth's rotation.

double star: A star that, upon closer or more sophisticated examination, turns out to consist of two or more component stars. The "primary" is the brighter of the two stars (although not necessarily the more massive and gravitationally controlling) and the other is called the "companion" or "secondary."

dust tail (of a comet): (see "tail").

eclipse: The hiding or dimming of one object by another object or by another object's shadow.

eclipsing binary: A type of variable star in which one component star of a double star eclipses the other, or both alternately eclipse each other, causing variations in brightness.

ecliptic: The apparent path of the Sun through the zodiac constellations, which is really the projection of Earth's orbit in the sky.

emission nebula: (see "diffuse nebula")

equatorial system: A system for indicating positions in the heavens using right ascension (corresponding to longitude on Earth) and declination (corresponding to latitude on Earth).

fireball: A meteor brighter than Venus.

full-cutoff fixture: A light fixture that emits light entirely below the horizontal, eliminating directly produced skyglow and reducing light pollution.

galactic cluster: (see "open cluster")

galaxy: An immense congregation of typically billions of stars forming a system of spiral, elliptical, or irregular shape.

gas giant: Planet of the type exemplified by Jupiter, Saturn, Uranus, and Neptune, consisting of a large gaseous part (huge atmosphere).

gas tail (of a comet): (see "tail").

geostationary (or geosynchronous) satellite: Satellite that orbits in the same amount of time it takes Earth to rotate and therefore remains in the same position with respect to any point of Earth.

gibbous: Phase in which the hemisphere of a world facing us is more than half lit but less than fully lit.

globular cluster: A kind of star cluster consisting of tens of thousands up to a few million stars arranged in a roughly spherical shape.

grazing occultation: Occultation in which one celestial object is only partly or intermittently hidden as it appears to move along the uneven edge of the closer object.

greatest brilliancy: Loosely speaking, the occurrence of maximum brightness of Venus.

greatest elongation: Maximum angular separation in the sky of one celestial object (usually an "inferior planet") from another (usually the Sun).

head (of a comet): A comet's solid "nucleus" and the "coma" cloud of gas and dust surrounding the nucleus.

horizon: The boundary line between sky and land or sea on Earth.

inferior conjunction: Arrangement in which an "inferior planet" passes the straight line between Sun and Earth.

inferior planet: Planet closer to the Sun than the Earth is to the Sun. The inferior planets are Mercury and Venus.

inner planet: One of the rocky worlds of the inner solar system, closer to the Sun than the asteroid belt. The inner planets are Mercury, Venus, Earth, and Mars.

Jovian planet: Jupiter and the planets that resemble it (see also "gas giant").

Kuiper Belt: A belt of comets starting at and extending far beyond the orbits of Neptune and Pluto, believed to be the source of "periodic comets."

libration: A slight nodding up and down and side to side of the Moon's Earthward face (usually visible over the course of weeks, and mostly caused by the Moon's variable speed in its elliptical orbit).

light pollution: Excessive or misdirected artificial outdoor lighting.

light-year: The distance that light, the fastest thing in the universe, travels in the course of one year.

limb: The edge of the Moon or other celestial body.

long-period comet: A comet that requires more than 200 years to orbit the Sun.

long-period variable: A major kind of variable star, in which the period of brightness variations is months or years long and the range of the variations typically great, with the amount of both often being irregular or only semi-regular.

magnitude: A measure of brightness in astronomy, in which an object one hundred times brighter than another is exactly five magnitudes brighter. The brighter the object, the lower the magnitude figure (e.g., a first-magnitude star is brighter than a second magnitude star), with negative magnitudes for the very brightest objects of all.

mare (plural **maria**): One of the vast gray plains of ancient lava on the Moon.

meridian: The line in the sky that passes from the due south horizon to the zenith onward to the due north horizon.

Messier objects (also known as **M-objects**): One of slightly more than one hundred deep-sky objects catalogued in the eighteenth century by French astronomer Charles Messier.

meteor: The streak of light seen when a piece of space rock or iron enters Earth's atmosphere and burns up (due to, most roughly put, friction).

meteorite: A piece of space rock or iron that survives its trip through the atmosphere as a meteor and reaches the ground.

meteoroid: A rocky or metallic natural object body smaller than an asteroid (no more than a few hundred meters across, usually very much smaller, even dust-sized) that would become a meteor if it entered Earth's atmosphere and a meteorite if it reached Earth's surface.

meteor shower: An increased number of meteors, all appearing to diverge from the direction of a single area among the constellations.

meteor storm: An extremely intense meteor shower, in which hundreds or even many thousands of meteors per hour may be observed.

meteor train: The glowing trail of ionization left by some meteors that lingers after the meteor itself has disappeared.

Milky Way: The great spiral galaxy that we live in, and also the night sky's band of strongest glow from the combined light of innumerable distant stars in the galaxy's equatorial plane.

moon: A rocky or icy object that circles a planet. Also known as a (natural) satellite.

multiple star: A star system consisting of more than two stars (although "double star" is often used as the umbrella term for systems of two, three, four, or more stars).

natural satellite: (see "satellite")

nebula (plural **nebulae**): A vast cloud of dust and gas in interstellar space. Different types include "diffuse nebula" (which include "emission nebula" and "reflection nebula"), "planetary nebula," and "dark nebula."

neutron star: The collapsed, ultra-dense core of a star left after a supernova, formed from an original star not massive enough to collapse all the way into becoming a "black hole."

nodes: The two places where a celestial object's orbit crosses the ecliptic or plane of Earth's orbit. The "ascending node" is the point where the object passes northward across the ecliptic (or, in three dimension, northward through the plane of Earth's orbit); the "descending node" is the point where the object passes southward across the ecliptic (or southward through the plane of Earth's orbit).

Northern Lights: (see "aurora")

nova (plural **novae**): An exploding (and therefore briefly very much brightened) star that loses a small fraction of its mass in the outburst, which may often arise from interactions between stars in double star systems.

nucleus: The solid, relatively permanent part of a comet, consisting of a strange assortment of dusty ices and perhaps some rock.

occultation: The hiding of one celestial object by another (sometimes such an event is instead, or at least primarily, called an "eclipse").

Oort Cloud: A vast but incredibly thinly scattered cloud of comet nuclei that extends from beyond the more flattened Kuiper Belt of comets to about halfway to the nearest star other than our own. The Oort Cloud is thought to be the source of long-period comets.

open cluster: This major kind of star cluster consists of a usually irregular shape and includes typically dozens or a few hundred stars (also called "galactic cluster").

opposition: The observationally very favorable presentation of a superior planet when it is 180 degrees from the Sun and thus rises approximately at sunset and sets at sunrise.

optical double: A double star in which the two components are not truly related, one object being much farther away and just happening to lie on nearly the same line of sight as seen from Earth.

outer planet: A planet in the outer solar system, farther from the Sun than the asteroid belt. The outer planets are Jupiter, Saturn, Uranus, Neptune, and Pluto.

P.A.: (see "position angle")

partial eclipse: An eclipse in which only part of a celestial object is hidden by another object, or only part of it is hidden by the second object's shadow.

penumbra: The lighter, outer, peripherial shadow of the Earth or other body.

penumbral eclipse: The least striking kind of lunar eclipse, one in which the Moon never enters the dark, central shadow (the "umbra") of Earth but instead only the lighter, outer shadow (the "penumbra") of Earth.

perigee: The near point on an orbit around the Earth.

perihelion: The near point on an orbit around the Sun.

periodic comet: Comet with an orbital period of less than two hundred years (also called "short-term comet").

phase: The amount of illuminated portion displayed by the Moon, a planet, or other celestial object that shines by reflected light.

planet: A relatively massive world (but not massive enough to sustain thermonuclear reactions like a star) in direct orbit about a star.

planetary nebula: A cloud of gas and dust cast off by a hot, small, dying white dwarf star. (The name comes from the passing resemblance of some of these blue or green nebulae to planets like Uranus and Neptune as seen in the telescope.)

position angle (P.A.): The direction angle of a point along the limb of the Sun or Moon, or of a "companion star" in relation to its "primary" in a double star system.

precession: A slight wobble in the rotational axis of the Earth, caused by the pulls of the outer solar system bodies, and resulting in slow changes of the direction of the north celestial pole and other positions in the heavens.

primary star: (see "double star").

prominences (solar): Immensely long streamers of gas temporarily levitated above the Sun's surface, apparently by forces in the Sun's magnetic field. They are sometimes directly visible—striking red in color—during a total eclipse of the Sun.

pulsar: A type of neutron star oriented toward Earth in such a way that we get to observe pulses of light and sometimes other electromagnetic wavelengths released from gaps in the exploding star's magnetic field near its poles.

quadrature: Position in which a planet (or other object) is located 90 degrees from the Sun in the sky, thus highest in the sky at either sunrise (west quadrature) or sunset (east quadrature).

quasar: An incredibly powerful source of light and other electromagnetic wavelengths that may be the intense core of a galaxy.

R.A.: (see "right ascension")

radiant: The region in the sky from which the meteors of a shower all appear to diverge.

rays (on the Moon): Streaks of light-colored material ejected from some of the younger craters on the Moon.

red giant: A huge and fairly massive star of extremely low density that radiates mostly in the red due to its relatively low surface temperature.

reflection nebula: (see "diffuse nebula")

retrograde motion: The less usual westward apparent motion of planets in relation to the background stars, an appearance caused by our vantage point as we catch up to a superior planet or an inferior planet catches up to Earth.

revolution: The orbiting of one celestial body around another (the Earth's revolution period around the Sun is one year).

right ascension (R.A.): West-east measure in the equatorial system of celestial coordinates, corresponding to longitude on Earth (though expressed somewhat differently than longitude on Earth: in hours of right ascension from 0 to 24).

rills: Narrow and sometimes meandering valleys on the Moon.

rotation: The spinning of a celestial object (the Earth's rotation period is one day).

satellite: Any celestial body that orbits another. In practice the term is usually confined to a body that orbits a planet. An "artificial satellite" is one launched by humankind to orbit Earth or another world; a "natural satellite" (composed of rock and/or ice) is more popularly known as a "moon."

secondary star: (see "double star").

"seeing": Sharpness of astronomical images as a function of turbulence in Earth's atmosphere.

short-term comet: (see "periodic comet")

sidereal time: Time measured by the passage of the stars around the sky, without reference to the Sun (the "sidereal day" is about four minutes shorter than the "solar day").

skyglow: The component of "light pollution" (excessive and misdirected artifial outdoor lighting) that goes up into the sky. Around even just fairly large cities, it is visible for dozens of miles.

SNR: (see "supernova remnant")

solar corona: (see "corona")

solar projection: Method of casting the Sun's image (with an optical instrument or mere "pinhole camera"–hole in cardboard).

solar system: The whole collection of planets, moons, asteroids, comets, and meteoroids orbiting the Sun under its gravitational influence.

sporadic meteor: A meteor not traceable to any known meteor shower.

star cluster: A grouping of anywhere from a few to several million stars traveling through space relatively close together but not closely enough to be considered a multiple star.

stationary point: The place at which the apparent movement of a planet in right ascension (or ecliptic longitude) stops and reverses due to a change from direct motion to retrograde motion, or vice versa (retrograde to direct).

superior conjunction: The position when a superior or inferior planet on the far side of the Sun passes the extension of the line from Earth to Sun.

superior planet: A planet farther from the Sun than the Earth is from the Sun. The superior planets are Mars, Jupiter, Saturn, Uranus, Neptune, and Pluto.

supernova: The very powerful kind of star explosion and brightening, in which a large part of a star's mass is lost, and the star's core may become a neutron star or black hole.

supernova remnant (SNR): The cloud of material ejected by a supernova, sometimes visible for many thousands of years after.

synodic period: The amount of time it takes for a planet to return to the same position relative to Earth and the Sun.

tail (of a comet): The long straight streamer (gas tail) or broad curved fan (dust tail) of material driven in the anti-sunward direction from a comet's head by the particles of the "solar wind" (in the case of the gas tail) and the radiation pressure of sunlight (in the case of the dust tail).

terminator: The line dividing day from night on a world.

terrestrial planet: A rocky, relatively small planet resembling Earth (including Earth) as opposed to Jupiter and the "Jovian" or "gas giant" planets. The terrestrial planets are Earth, Venus, Mars, and Mercury.

total eclipse: Eclipse in which the Sun's blinding surface is completely covered by the Moon or the Moon is completely covered by Earth's "umbra" (central shadow).

transit: The apparent passage of one celestial body in front of the face of a much larger one as seen from a third location—there are transits of Mercury and Venus in front of the Sun, and of the Galilean satellites in front of Jupiter.

transparency: The degree to which celestial light is able to pass through Earth's atmosphere.

umbra: The central shadow of the Earth (or other object).

Universal Time (UT): Time system for dating astronomical events, corresponding to the local time at Greenwich, England, on the 0 degree meridian of longitude on Earth (to obtain Universal Time from your current local standard time, subtract five hours from Eastern Standard Time, four hours from Central Standard Time, and so on.)

UT: (see "Universal Time").

variable star: A star that for one of a number of possible reasons undergoes changes in its brightness.

waning: A decreasingly illuminated phase.

waxing: An increasingly illuminated phase.

zenith: The overhead point in the sky.

zodiac: The circle of constellations through which the Sun passes during the course of the year.

zones: The lighter-colored bands of clouds on the face of Jupiter and Saturn.

Sources of Information

✦ ✦ ✦

BOOKS AND HANDBOOKS

Burnham, Jr., Robert. *Burnham's Celestial Handbook* (3 volumes). New York: Dover Publications, 1983. The Classic and still unrivaled twentieth-century guide to most deep-sky sights visible through small or medium-size (up to about ten-inch aperture) telescopes. A grand collection of information and photographs, with lucid text about both the lore and the science of thousands of stars, constellations, and other objects.

Covington, Michael. *Astrophotography for the Amateur.* New York: Cambridge University Press, 1991. A good introduction to astrophotography.

Dickinson, Terence, and Alan Dyer. *The Backyard Astronomer's Guide.* Camden East, Canada: Camden House Publishing, 1991. Insightful and comprehensive coverage of what amateur astronomers need to know about equipment, techniques, and observing sites. Available from Camden House Publishing, 7 Queen Victoria Road, Camden East, ON K0K 1 J0, Canada.

Harrington, Phil. *Touring the Universe Through Binoculars.* New York: John Wiley & Sons, 1990. Excellent guide to selection and use of binoculars. The majority of the book is devoted to constellation-by-constellation tour of binocular sights.

Ottewell, Guy. *The Astronomical Companion.* Greenville, SC: Universal Workshop, 1979. Ottewell's 72-page atlas-sized guide to non-year-dependent astronomical topics (see his "Astronomical Calendar" below for those that are), including dozens of huge and unique diagrams. Available from Universal Workshop, Furman University, Greenville, SC 29613.

Raymo, Chet. *365 Starry Nights.* New York: Simon & Schuster, 1992. Superbly written and charmingly illustrated night-by-night guide to seeing and learning the heavens.

Schaaf, Fred. *Seeing the Sky, Seeing the Solar System,* and *Seeing the Deep Sky.* New York: John Wiley & Sons, 1990, 1991, 1992 respectively. Trilogy of books of observing projects focusing on naked-eye observations, telescopic observations of solar system objects, and telescopic observations of objects beyond the solar system.

Schaaf, Fred. *Wonders of the Sky,* New York: Dover Publications, 1983. Introductory guide to naked-eye skywatching.

ATLASES

Star atlases range from those displaying only stars down to magnitude 5.0 or 6.0 (a few thousand stars) to *The Millenium Star Atlas* (well over one million stars, almost all stars brighter than magnitude 10.5). A fine selection of these atlases is described in the Sky Publishing Corporation catalog (see address and phone number for *Sky & Telescope* magazine below).

Various sky simulation software programs offer access to computer maps of even dimmer stars. Descriptions of many of these can also be found in the Sky Publishing Corporation catalog and elsewhere (catalogs of the Astronomical Society of the Pacific, and of the Orion Telescope and Binocular Center, for instance).

For lunar observers, a beautiful and most informative volume, *Atlas of the Moon* by Antonin Rukl, is available from Kalmbach Publishing (see address for *Astronomy Magazine* below).

PERIODICALS, ALMANACS, ANNUAL GUIDES

Astronomical Calendar. Seventy-two atlas-sized pages filled with original diagrams and rich text about the year's celestial events (includes several sections and month-by-month "Observers' Highlights" by the author of the book you are holding). By Guy Ottewell at Universal Workshop (see address above).

Astronomy Magazine. Kalmbach Publishing, P.O. Box 1612, Waukesha, WI 53187.

Mercury. Magazine published by Astronomical Society of the Pacific (see address under "Audiovisual Materials and Computer Software" below).

Sky & Telescope. Includes planets and stars columns (plus other pieces) by the author of the book you are holding. Subscription orders by phone: 800-253-0245, or write to: P.O. Box 9111, Belmont, MA 02178.

Sky Calendar. For just $9 a year you get a two-sided sheet for each month with a basic star map on one side and an informative calendar with a sky scene for each night on the other side. Write to: Abrams Planetarium, Michigan State University, East Lansing, MI 48824.

PHONE HOTLINE

"Skyline": 617-497-4168. A service of *Sky & Telescope,* this is a recorded phone message about current news and sights in astronomy. The message is updated each Friday, or sometimes more frequently as circumstances dictate. (A version of the "Skyline" text, usually with some illustrations or additional information, is presented online as "S&T News Bulletin"—available in America OnLine's Astronomy section or from SKY Online, listed below.)

ONLINE SOURCES

American Association of Variable Star Observers: http://www.aavso.org/

Association of Lunar and Planetary Observers: http://www.lpl.arizona.edu/alpo/

Astronomical League (confederation of more than two hundred amateur astronomy clubs in the U.S.): http://www.mcs.net/~bstevens/al

Astronomical Society of the Pacific: http://www.aspsky,org/

Comet Obversation Home Page: http://encke.jpl.nasa.gov/index.html

International Dark-Sky Association (central bureau for information on light pollution and how to combat it, technologically and legally): http://www.darksky.org/

International Meteor Organization: http://www.imo.net/

International Occultation Timing Association: http.www.sky.net/~robinson/iotandx.htm *and* dunham@erols.com

SKY Online (home page of *Sky & Telescope* magazine and the Sky Publishing Corporation): http.www.skypub.com

AUDIOVISUAL MATERIALS AND COMPUTER SOFTWARE

Astronomical Society of the Pacific. 390 Ashton Ave., San Francisco, CA 94112. 415-337-1100. This century-old educational organization offers a catalog with a wide variety of excellent books, posters, slide sets, computer programs, and CD-ROMs.

Sky Publishing Corporation. P.O. Box 9111, Belmont, MA 02178. 800-253-0245. E-mail: orders@skypub.com. The publishers of *Sky & Telescope* offer a catalog with books, star atlases, computer programs, CD-ROMs, slide sets, posters, and even globes (including ones of other planets).

INDEX

✦　✦　✦